Pro Tools 8

Pro Tools 8

Pro Tools for Film and Video

Dale Angell

Routledge
Taylor & Francis Group

LONDON AND NEW YORK

First published 2009

This edition published 2015 by Focal Press

Published 2017 by Routledge
2 Park Square, Milton Park, Abingdon, Oxon OX14 4RN
711 Third Avenue, New York, NY 10017

First issued in hardback 2017

Routledge is an imprint of the Taylor and Francis Group, an informa business

Library of Congress Cataloging-in-Publication Data

Angell, Dale.
    Pro tools for film and video/Dale Angell.
        p. cm.
Includes bibliographical references and index.
ISBN 978-0-240-52077-3 (pbk. : alk. paper)
    1. Digital cinematography. 2. Motion pictures--Editing--Data processing. 3. Pro tools. I. Title.
TR899.A586 2009
778.5'3--dc22

                                                                2009009654

British Library Cataloguing-in-Publication Data

A catalogue record for this book is available from the British Library.

ISBN 13: 978-1-138-46840-5 (hbk)
ISBN 13: 978-0-240-52077-3 (pbk)

Typeset by: diacriTech, Chennai, India

Dedication

Figure 1 Ruth Louise Edmonds Angell Grose Newbold, AKA Mom at the keys of her M-audio system and iMac.

This book is dedicated to my mother, who not only brought me life but taught me to enjoy it. She put up with, and funded, my weird desire to build a recording studio in the garage (complete with pipe organ) and put up with calls from the neighbors complaining about strains of Toccata and Fugue in D minor in the middle of the night.

She helped me infest my history teacher's yard with gophers, turned a blind eye to my creative use of toilet paper on the neighbors, houses, and put up with both my brother and me dragging home some of the weirdest stuff (and creatures).

She has given me her love of music and I have returned the favor by turning her into a Mac nerd. Thanks, Mom.

Contents

Contents

Contents

Contents

Introduction

Figure 1 Dr. Tom Stockham at the University of Utah 1976. Image courtesy of the Marriott Library special collections.

Pro Tools was released in 1987 as the Mac-based "Sound Tools." Until that time, digital "tapeless" systems were very experimental and proprietary. In 1976 Dr. Tom Stockham of the University of Utah made the first commercially successful 16-bit digital recording on his "Soundstream" digital recorder. The recorder was a reel-to-reel tape system using one-inch data tape. Because no consumer digital format existed at the time, these recordings were sold as vinyl records on the Telarc label. Dr. Stockham also edited his recordings on a fully digital, nonlinear, hard-drive-based Digital Equipment Corporation PDP 11/60. Today we refer to this type of editing system as a Digital Audio Workstation or DAW.

In 1979 the DAW found its way into the cinema when Neiman-Tillar Associates developed what they called the Automated Computer Controlled Editing Sound System (ACCESS). This system was able to interlock the DAW to picture and interface with the analog mixing systems in use at the time. They received the Academy Award in 1979 for the creative development of this system. The system was proprietary to Neiman-Tillar and only used in-house for sound effects and music editing for motion pictures.

Two years later, in 1981, several major technical developments again changed the way audio was recorded, distributed, and edited. Philips demonstrated the Compact Disc or CD. While this was still several years away from consumer shelves at a price that the average person could afford, it introduced the public to the "digital sound." Interestingly, the initial public reaction was dislike. There were complaints of "hard-edged sound," "lacking warmth," and "digital flu." (Digital flu???) Soon the format found wide acceptance and, as they say, the rest is history. That same year IBM released the first Personal Computer (PC).

Also that year the MIDI format was released. MIDI or Musical Instrument Digital Interchange was intended to interface synthesizers and other keyboards to each other. Perhaps it is no coincidence that MIDI and the PC came out at the exact same time; from the first moment, the idea of interfacing the PC into the MIDI control interface was obvious.

Figure 2 The original Mac 128.

In 1984 Apple released the Macintosh. Critics called it the "toy computer." After all, it looked like a toy. However, the price tag was not very toy-like, but then neither was the performance.

Figure 3 The Mac 2b.

In 1987 when Sound Tools was released it ran on the Macintosh II, a 128 K (expandable to 20 meg) 16-megahertz version of the "toy computer." What made it possible to edit audio was not the horrendous speed of the processor or the awesome amount of RAM memory; this was the first Mac to offer NuBus expansion slots. This made it possible to interface the Mac Plus with outboard interfaces, and save files on a hard drive. Hard drives were available in sizes including 20, 40, 60, and even massive 80 meg. Keep in mind that that is about 1/10 the size of an audio CD. Keep in mind also that the power of this computer was about 1/5000 of what we expect today.

Nevertheless, it worked. It even worked well. In 1991 the name was changed to Pro Tools and DSP was added to the NuBus creating the basic TDM system. It soon became the dominant force in audio recording, film and video sound editing, and mixing. In 1997 Avid, makers of the leading film and video editing software, purchased Digidesign, the makers of Pro Tools, which by then had become the predominant film and television sound editing and mixing system.

While both the Avid and Pro Tools were Mac-based in their original versions, both were soon offered as PC-based systems. In fact, something like a war broke out between Apple and Avid. Apple released Final Cut Pro, a video/film editing system intended to challenge Avid's position in the film industry. While Avid is still the predominant picture editing software, Final Cut Pro has cut into that market in a big way. And Apple also seems to want to release an Apple version of everything Avid offers.

The good news for sound people is that this "war" has driven the prices down – in fact right through the floor. With new technology constantly becoming available, and with this competition nipping at their heels, the capabilities of the systems are now phenomenal. And while it is still possible to spend over $100000 on a Pro Tools system, it is also possible to add an M box and Pro Tools software to your laptop for about $500. And it isn't just Apple and Digidesign fighting over this market; scores of other DAW systems are now out there, each with their own loyal users.

But nothing on the market challenges Pro Tools. It is simply the de facto audio system in music recording, film, and television. In 1999 Tom Stockham received an Academy Award for the development of digital audio recording and editing. He had also received a 1988 Emmy and the first ever technical Grammy Award in 1994. He is one of the few people to have awards in all three areas. And in 2004 Digidesign received an Academy Award for the development of Pro Tools.

The Kodak 35mm Project – "Loves Devotion Forever"

Figure 1 "Loves Devotion Forever" was the 2006 Kodak 35mm project by Brooks Institute's school of film and television. Over 150 Brooks students worked side by side with industry professionals making this 20 minute film. Photo by Pedro Gutierrez.

The Kodak 35mm project is a program of Kodak's Student Filmmaker Program. The project gives student filmmakers the opportunity to create a 35mm motion picture using the same tools used by filmmaking professionals. Kodak sponsors The 35mm Project in conjunction with Mole Richardson, Clairmont Camera, FotoKem Laboratories, Dolby Laboratories, Laser Pacific, Mix Magic, NT Audio, and FPC.

The sponsors provide 35mm motion picture film, film processing, camera package, lighting, grip and generator equipment, surround sound and audio mixing, front and end titles, optical sound transfers to film, telecine, and answer printing. Students are also mentored by top filmmakers, including members of the American Society of Cinematographers like Laszlo Kovacs, ASC and Richard Crudo, ASC, various members of the American Cinema Editors, and professional sound mixers. The sponsors also provide workshops and demonstrations on the use of cameras and equipment.

The concept for The 35mm Project was developed in 2001 by Lorette Bayle of Kodak. The pilot schools included Chapman University and the California Institute for the Arts. The 35mm Project was extended to include the University of California Los Angeles, Scottsdale Community College, Loyola Marymount University, California State University of Long Beach, University of Arizona, California State University Northridge, the University of California Santa Barbara, and Brooks Institute.

Figure 2 Cinematographer Chuck Minsky, seen here at the Santa Barbara pier, worked with student director of photography Mat Walla and director Jessie Haggy to achieve the look of the film. Many industry professionals shared their talents with students on the three-day shoot. Photo by Jacob Foko.

Figure 3 Mentors worked with students in all departments. Left to right – directing mentor Perry Lang (*Everwood, NYPD Blue, Medium*) student producer Alex Zhan, producer mentor Steve Tracxler, (*Windtalkers, Legally Blond 2, Out of Time*) and director of photography mentor Chuck Minsky (*You, Me and Dupree, The Producers, Pretty Woman*). Photo by Pedro Gutierrez.

The 60-year-old Brooks Institute in Santa Barbara California has become a regular participant in the 35 mm project. Brooks Institute graduates are visible nationally and internationally, working for distinguished organizations including National Geographic, Smithsonian, the Los Angeles Times, and other national media outlets, including Hallmark Publishing, Cousteau Society, HBO, Kodak, and other industry leaders in visual media fields. Brooks has brought in several sponsors to support their annual project including The Santa Barbara International Film Festival, Fisher Light, T and T effects, Match Frame Video, and Chapman Leonard.

Brooks professors Tracy Trotter and Dale Angell enlist the help of a who's who list of industry experts. Students work under the mentorship of professionals including producer Steve Traxler whose credits include *Legally Blonde* and *Out of Time*; actor John Cleese of Monty Python's Flying Circus; director of photography, Chuck Minsky, DP on *Pretty Woman, Almost Famous*, and *The*

Producers; Bill Butler, Director of Photography on over 80 films including *Jaws, The Conversation,* and *Biloxi Blues;* director Perry Lang of *Alias, Dawson's Creek,* and *NYPD Blue;* and photographer Wayne Goldwyn, whose credits include *CSI, Jeopardy,* and *Nixon.* The list of professional mentors numbers in the dozens and changes every year.

Figure 4 Production mixer mentor Bruce Pearlman works with student Matt Sanchez. Photo by Pedro Gutierrez.

The Kodak 35mm Project at Brooks involves a 150-person film crew, more than 30 vendors, rental equipment valued at more than $1 million, and an estimated 24 000 hours of labor from students, faculty, and volunteers. Photo by Pedro Gutierrez.

"Loves Devotion Forever"

Directed by
Jessie Haggy

Written by Jessica Kalin

"Loves Devotion Forever" was the third student film made by students at the Brooks Institute as part of the Kodak student filmmaker program. It was shot and finished on Kodak 35mm with a Dolby LCRS and Dolby 5.1 soundtrack.

Figure 5 Sound cart on stage 2 for the house scenes. Production audio was recorded on a Audio Devices 744T digital recorder with smart slate The edit was performed on Final Cut Pro/Cinema Tools with Pro Tools LE and HD 7.3 for sound design and mixes. Photo by Pedro Gutierrez.

We will be following this film in each chapter as it moves through its workflow. Like all films, the project started with a script, in this case a submission of student scripts. Certain elements were required to be in every script submitted; each project had to have be a romantic comedy, have a scene at a party, and contain a falling stunt from the Santa Barbara pier.

Figure 6 The set for Madeline's house is a "left over" from the film *Erin Brockovich*, which was filmed on stage two in 1999. Photo by Pedro Gutierrez.

A crew was selected from resumes and reels submitted by students who were vying for one of the forty coveted major crew positions. The sets were designed and constructed, costumes and props located, and the production schedule put together. Student producers arranged insurance, permits, equipment deals, housing for mentors, and a thousand and two last minute changes.

The shooting schedule was three days. Everything from the equipment to the schedule and the strange requirements on the script were meant to create a real model of a Hollywood feature. All union requirements were also met whenever possible. The project truly gave the students knowledge of what goes on when making a modest budget feature, plus real world on set experience.

Figure 7 Student camera department working with the Arri Moviecam at the Brooks Institute. Photo by Pedro Gutierrez.

The students also attended various workshops in Los Angeles on the camera systems to be used, lighting systems, generators, sound gear, and support systems. They were also guided in scheduling, getting insurance, working with municipalities gaining permits, and arranging transport and housing.

Most of the student crewmembers had an industry mentor working directly with them during production. The project was shot in conjunction with the Santa Barbara International Film Festival; several scenes were shot during festival events and were open to the festival attendees.

Figure 8 many of the extras in the theater scene were Biggins inflatable dummies. Blowup extras have been used successfully on scores of films. *Seabiscuit* used 7000 of these blowup people. Photo by Pedro Gutierrez.

About Pro Tools 8

Pro Tools 8

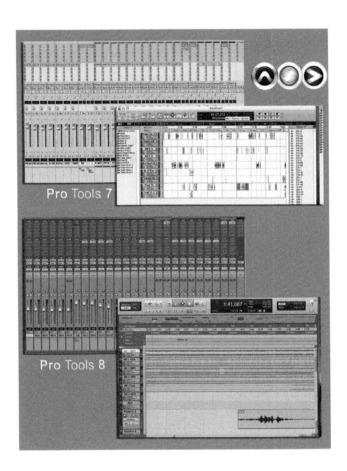

Pro Tools 8 brings some brilliant new features to Pro Tools, and while most benefit the MIDI and music guys (or it sure always seems like that), there are many new tools for postproduction as well. In the LE and M-Powered versions, you now have up to 48 mono or stereo tracks, and the long awaited 5.1 mixing in the LE version. The cost for this add-on is high; it requires both the DV Toolkit 2 software that adds timecode clock, DigiTranslator, TL Space convolution reverb, VocALign Project, and BNR noise reduction as well as the Music Production Toolkit that adds the Digidesign Hybrid synthesizer, SoundReplacer, and Smack! compressor. With the new Complete Production Toolkit (Pro Tools LE 8 only), which combines both toolkits, you can also work with up to 128 audio tracks with 7.1 surround mixing. This would also require the use of an audio interface with at least eight in and outputs, such as the 003. But now for around $3000 you have an LE system that (almost) rivals the HD systems.

Pro Tools 8 also has a new feature called Satellite Link. This is a simple linking system that allows up to five HD systems to be connected without the need for timecode chase. And in the LE version, Satellite Link is used to interlock with a dedicated video system that offers many of the features of Virtual VTR (Chapters 1–10). This is great as it frees up the processor for more audio, and the video player can play Avid and QuickTime HD/SD video.

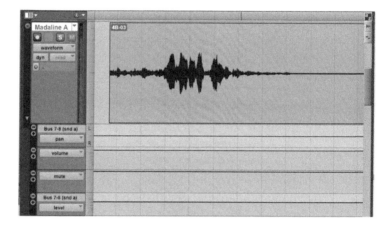

Figure 2 Pro Tools 8 now allows MIDI CC and automation to be viewed and edited in lanes or in the original overlay display.

The look and feel of the interface is totally new, yet the tools are the same, and in the same places. Unless you move them somewhere else, that is a new feature: a customizable interface. Those pesky little buttons that you need a magnifying glass to see, like the Nudge control and Link selection, are now big and easy for old geezers (like this author) to see without getting whiplash from our bifocals. You can change the colors of the markers and channel strips to any color you want, and there is a great new feature in the Edit window where you can now display all of the automation in "Automation and Controller lanes" to view and edit audio track automation (volume, pan, and plug-in automation) or for the MIDI guys, MIDI CC data (velocity, pitchbend, and modulation) without changing track views.

There are now 10 inserts per track, and a new "universe" window, which looks like a track at the top of the edit window but is in fact a map to the entire session with a sort of "you are here" square showing the currently displayed area. This makes it possible to jump all over the session like Spock beaming in from the starship. Very fast, very cool.

There is a new track comping feature. Often in music production different takes are recorded to different "playlists" in the same track (HD systems only). Any of these takes can be selected for playback on that track. But with the new comp tool, small sections of each playlist (take) can be selected for playback at the current time. This is great for music tracks, but it could be useful in ADR and Foley recording as well (Chapters 5 and 6).

But wait, there's more! I've saved the best for last (that is to say, if you are a MIDI/music person)! There are some fabulous new capabilities for music production. The sessions and regions are now totally elastic; each region can be sped up, slowed down, matched to the session speed, the beats aligned, and each note can be pitched up or down into tune. This also makes it possible to transpose to new keys, and to match speed and rhythm to MIDI tracks.

There is also a studio full of virtual instruments including the new Mini Grand piano, Boom drum machine, DB-33 tonewheel organ (gee, I wonder how they came up with that name?), Vacuum and Xpand synths, and the new Eleven Free and SansAmp guitar amps. There's also DJ software and hours of loops to make your own compositions. You can also now display MIDI tracks as standard MIDI tracks or as notation. Data can be written directly on the notation tracks using the Score Editor or in MIDI using the MIDI editor.

Now, if all of this is sounding a bit overwhelming for we post people, there is an act of revenge. No one can stop you from playing with their toys. And they are very fun to play with.

Where necessary, this book shows both the Pro Tools 7 LE interface and the Pro Tools 8 LE interface. There are a few tricks to overcome shortcomings in Pro Tools 7 LE or Pro Tools 8 M and 8 LE when not using the new Complete Production Toolkit for 8 LE.

About the Web site

The Web site contains the entire sound edit for "Loves Devotion Forever". We will be using this sound edit and mix as an example of the techniques covered in each chapter. The project files require Pro Tools version 7.0 or greater and use QuickTime playback. They can be opened on M, LE, or HD systems.

The projects on the Web site are the OMF import, the final M and E (music and effects stems), and final delivery to the mix stage on the Brooks/Kodak 35mm project "Loves Devotion Forever". The projects are compressed to 44.1 K 16-bit audio and highly compressed QuickTime video. These settings are for the demo project only, and these settings would never be used on an actual film/video project. The actual project was 48 K 24 bit with a DV format QuickTime video.

Because the video is compressed, and no longer in DV format, it will not play to the video monitor through a DV firewire device but it must be played on the computer monitor. If you have three computer monitors or a wall-sized LCD screen, you will find this expectable; the rest of the world will feel rather cramped.

The OMF import is how the project was delivered from the picture editor. The OMF file was opened with Digitranslator. This step has already been done for those without this expensive plug-in.

As the Pro Tools session cannot be played directly from the Web site, copy the entire Pro Tools Session folder to your hard drive. For those using Pro Tools 7 LE, you will find that there are too many tracks to playback. If you make temp music tracks three through six inactive, this should solve the problem. And as these tracks are included only so that you can switch to the original temp music to compare with the final score, making these muted tracks inactive will not alter the final mix. Temp music one and two contain some temp score that was retained in the final mix and so are needed to hear the entire mix. To listen to the entire temp score, simply make all the final score tracks inactive and switch off automation for the temp score tracks so that they don't auto mute or temporarily delete the mute automation for these tracks.

In this chapter

Pro Tools systems

1.1 The audio interface: M-Powered, LE, and HD

All Pro Tools systems require some kind of hardware audio interface. When working with a video reference, there also must be some system to play back the video and maintain accurate sync with the Pro Tools session. Some of these interfaces are very simple, some are rather complex, but all serve the same functions.

The audio hardware interface must be matched to the version of software being used. Digidesign currently offers three versions of their software, M-Powered, LE, and HD.

Figure 1.1 Pro Tools M-Powered.

The M-Powered software is used with the line of M-Audio interface devices. M-Audio also offers a unique line of software that is compatible with their interfaces and a variety of hardware and software interfaces. Some of the M-Audio interface devices, for example, their Musical Instrument Digital Interface (MIDI) interfaces, will work with any version of the Pro Tools software.

Generally, M-Audio is its own unique system that can also use Pro Tools M-Powered software.

Pro Tools compatible M-Audio interfaces include small USB and FireWire boxes with mic inputs, instrument inputs, Sony/Philips Digital Interface (S/PDIF) and Alesis Digital Audio Tape (ADAT). Some feature HD sample rates and 24 bit depth. There is even a I/O/control surface combo similar to the Digidesign 003. They also supply audio cards with digital signal processing (DSP) and rack systems. M-Audio also manufactures a line of MIDI interfaces and high-quality keyboards that work well with the new Pro Tools virtual instruments with HD or LE software.

Figure 1.2 Pro Tools LE.

The LE systems are an extremely powerful "down-scaled" version of Pro Tools HD. Most of the functionality of Pro Tools is present in the LE version. LE and M-Powered sessions are also compatible with the HD and older MIX systems. Generally speaking, any version 7.2 or newer Pro Tools session can be opened on any Pro Tools system if the sample rate is supported. While some of the advanced functions used in an HD session lock out when this session is opened on an LE system, they are not lost. All functions come back online as soon as the session is opened back on the HD system. When opening a project made with newer software in older software versions, there may be some compatibility problems. While all projects going back to version 3.2 can be opened in versions 7.2 through 8.0, any version 7.0 or 8.0 project to be opened on an older version 6.9 to 3.2 must first be saved by performing a "Save copy in" and setting the project version to the proper older version. While this will open the newer project on the older system, many features are lost and the newer plug-ins are not compatible. If you have a pre-7.2 system, God bless you, but really, it's time to upgrade.

This is great for sound editors. They can have an LE system on their laptop computer and still open the entire session. Audio can be recorded and edited anywhere. If you want to record Foley in a cave, this is not a problem. You'll

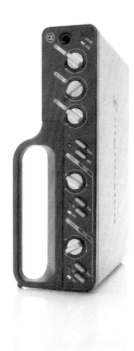

Figure 1.3 Pro Tools
Mbox 2.

have more trouble convincing the Foley walker to go into the cave than you will have setting up a full function Pro Tools system in there.

The LE software uses several audio interfaces. The Mbox 2 (not to be confused with M-Audio interfaces) has two analog audio inputs with mic preamps, phantom power, instrument input for electric guitars and other pickup equipped instruments plus line inputs. Two analog outputs, two digital S/PDIF in and outs and MIDI in and out.

The 003 and 003 Rack have eight analog inputs and outputs as well as eight mic preamplifiers. It also has eight digital ADAT inputs and two S/PDIF inputs and outputs. The 003 comes with an eight-fader control surface. As these systems have eight analog outputs, when using Pro Tools 8.0 LE with the Complete Production Tool Kit, the 003 can now be used for 5.1 surround mixing.

The HD systems are the flagship of Pro Tools. They can be configured to just about any input/output configuration imaginable. If you want to play 200 tracks out to 98 analog outputs, it's doable. The HD systems require several hardware interfaces. One to seven expansion cards are installed in the computer and/or an expansion chassis, and these connect to one or more I/O

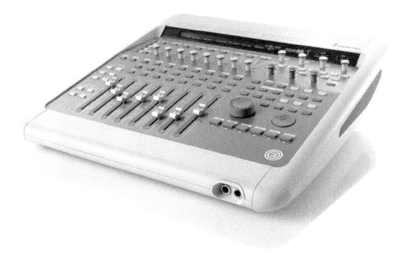

Figure 1.4 Digidesign
003 for LE systems.

Figure 1.5 Pro Tools HD.

interfaces. The I/O interfaces have eight analog and up to 16 digital inputs and outputs each. There is also an eight-channel mic preamp and a ten in and out MIDI interface that are separate.

The older MIX-series systems use this same system, but are no longer supported by Digidesign.

MIDI was originally intended to interface keyboards and other instruments to sample players and controllers. It has expanded in scope and is now used for a variety of purposes including device control, synchronizing video and making sound effects. Even if you have no interest in playing music, you may need a MIDI interface.

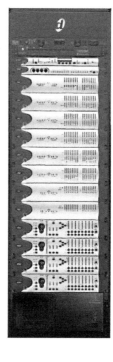

Figure 1.6 Rack of Pro Tools HD IO devices. In this case, six 192 I/O for 96 channels of digital I/O, two 96i I/O for a total of 128 analog audio I/O and 32 mic inputs from four PRE mic preamps. The rack also has MIDI I/O and SYNC HD, power conditioning and an expansion chassis for up to seven HD PCIE cards.

1.2 The video interface

Because we are editing sound to picture, we will also need some way to interlock the picture and hold it in accurate sync to the sound tracks. There are several systems being used to interface video to Pro Tools; each has advantages and disadvantages.

Video playback from QuickTime

It is possible to avoid using any machine control and simply play the reference from a QuickTime file. There are many plusses to using this system. It requires no extra hardware or software, and it can play at many frame rates, great for working with Internet media or 24 fps film projects.

The only down side is that it ties up CPU power and may mean using fewer plug-ins or even audio tracks. It also requires that the project be delivered as a QuickTime or that you have a system to capture QuickTime from video.

A great way to help with both problems is to have either a FireWire video IO device or a good FireWire DV recorder. For example, the Canopus ADVC-110 has video in and out directly to FireWire. The QuickTime movie can be set to "QuickTime Movie Out FireWire" and the ADVC-100 will convert this to video, which can be sent to an National Television System Committee (NTSC) monitor. Anytime the video is played out to the FireWire, CPU usage is greatly reduced.

There are also FireWire-to-video boards that can be installed in the computer. There are even ones for laptop computers.

Any DV recorder, even a DV camera, can be used as the FireWire video device; however, this does not mean that it can be used as the DV QuickTime capture device. Many DV devices capture with timecode and at proper speed, but some do not and should be avoided. Any DV Cam or DVC Pro device is fine, but a simple DV camera may be trouble. You will also need capture software; Final Cut Pro, Avid DV or Express, Adobe Premiere or even iMovie will work fine. Anything that can capture the DV to a QuickTime movie will work. To import the QuickTime movie, choose File > Import > Video.

When using the Video Movie Out FireWire function, the decoding video device delays the video. This knocks the session out of sync. To compensate for this, it is necessary to set a "Movie Sync Offset." This delays the Pro Tools audio by a preset amount bringing the session back into sync. Depending on the device, this delay may be several frames. The setting is made in quarter-frame increments. There are published numbers for some devices, however, these may not be correct on your system. Most people experiment to dial this in. The Video Sync Offset is in the Setup menu. Unfortunately, you can't simply stop on the two pop and line this up. There is no delay until playback starts. Use some dialogue and experiment. You will find a range of settings that seems to work. Set the offset to the center of this range.

1.3 Machine control and the SYNC HD

The oldest system, which is still widely used, is called "machine control." The very oldest film interlock systems locked picture and sound together with a bicycle chain driven by one huge motor out in the alley. The chain ran around the facility on pulleys. When you wanted to interlock one thing to another, you simply engaged its clutch to the chain sprocket and controlled the entire interlock with the motor controls. It sort of gives new meaning to the expression "gearing up for a session."

This is the basic idea behind machine control. When the Pro Tools system moves forward or backward, a video recorder and/or other equipment is forced to move with it – except a Pro Tools system doesn't have any moving parts where sprockets can be attached. Video recorders need to be controlled by something that is a bit more sympathetic to their needs than a bicycle chain.

Figure 1.7 House sync or the latest innovation from the Tour de France? Bicycle chains were once used to interlock recorders, playback machines (dubbers), and projectors.

But the chain idea is still a good one: a single reference that every piece of equipment hooks to and uses as a speed reference. There are many things that can be used, but the key is that everything needs to use the same reference. Two references that are very close are not close enough.

Most large facilities have a "house sync generator." The house sync is sent on video cable through the entire facility and can be used as a speed reference by any piece of equipment. Unlike the chain, it is always moving forward and never stops, but it also is much quieter and never needs oiling. The house sync is simply a black video signal at 29.97 fps (NTSC systems) or 25 fps (PAL systems) being created by a "black burst generator" or other type of sync generator that can supply black burst. There are several other speed references that can be used in a Pro Tools session:

- Digital audio with word clock. Any time a digital audio signal is recorded as a digital data stream, the recorder can lock to the incoming digital audio and use this as its clock. Audio Engineering Society (AES), S/PDIF, and ADAT are common.

- The computer's processor clock

- Line or pilot (50, 59.94, or 60 cycle reference, often from the line power or from a Nagra pilot. This requires Pro Tools HD software and a SYNC HD)

Unlike the bicycle chain, these only supply the speed reference. We also need a system to pull the elements into position (positional reference). On the old chain-driven system, this was done by placing a mark on every element at the "picture start" at the head of each reel. The loader simply lined up the marks and engaged the clutch. When the motor started, everything moved together in sync.

On modern systems, we use timecode for positional reference. If picture start is at timecode 01:00:00:00 on every element, the automated controller can simply line everything up and wait for a control command. If one element is not in line with this timecode, say its picture start is at 01:00:01:00, then this "offset" can be entered and the controller will hold it to this offset.

This means we have two signals that are used together to hold sync, the positional reference, usually timecode, and the speed reference. When setting up a Pro Tools system, both need to be configured.

There are several other kinds of positional reference signals:

- MIDI

- Bi-phase and Tach (used by film recorders and projectors)

While any Pro Tools system can use the computer's clock as the speed reference, Pro Tools LE cannot be locked to any sync other than the computer's clock or incoming digital audio and is not able to "resolve" speeds or use an external synchronizer. It does have a timecode trigger that can start playback of a video recorder at the same timecode of the Pro Tools session; however, there is no "bicycle chain" in this system. Each device is simply being put into play at the same time but then locking to its own speed reference. Unless they are locked to the same speed reference, within minutes they will drift apart, losing sync. When locking Pro Tools to a system or device with digital audio out, for example, S/PDIF, if Pro Tools uses the S/PDIF as the clock, then sync can be held indefinitely. When locking to digital audio output, timecode trigger is not just the best system, it is the only system.

LE systems can lock to other devices as long as those devices also lock to the computer's clock. Because LE systems use the computer's processor as their sync reference, the processor becomes the speed reference for the bicycle chain. MIDI, Ethernet controlled hardware, and QuickTime can all be locked to the computer's processor clock and therefore locked to Pro Tools LE systems. This allows syncing QuickTime to Pro Tools without sync drift problems. As always, the key is locking everything to the same reference.

Using the SYNC HD (or older USD)

The SYNC HD allows a Pro Tools HD system to "resolve" to various clock sources and positional references. Resolving refers to controlling a system's speed to keep it in sync with other devices or in sync to a reference. The SYNC

Figure 1.8 The Pro Tools SYNC HD is used to lock Pro Tools to other equipment while holding sync.

HD is a controller capable of resolving many different systems. One of the pieces of interfaced equipment becomes the "master," and the rest resolve to it. And the master device is not free-running; it too is resolved to the speed reference. It is only the master in terms of positional reference.

Video playback via machine control

To use machine control video on Pro Tools, a Pro Tools software option must be installed. You will also need a synchronizer – there are several third-party devices that will work, but the gold standard is the Digidesign SYNC HD.

Video recorders capable of machine control have a 9-pin remote control capability. When in the remote setting, the front panel controls are disabled and control of the recorder is from the 9-pin connector on the back. Such video recorders also have a time base corrector (TBC) and are capable of resolving their speed to house sync.

Pro Tools controls the recorder with both software and hardware. When the SYNC HD is locked to house sync, and Pro Tools is locked to the SYNC HD, it will play at the exact same speed as the video recorder. But it also needs to have a timecode reference to pull it into the proper position before releasing both systems to play at speed.

The timecode from the tape recorder is recorded onto the video tape. On some formats, this is recorded along the edge of the videotape as a linear, longitudinal track. This is referred to as Longitudinal Timecode (LTC). Other video formats use timecode recorded into the vertical sync of the video signal. This is referred to as Vertical Interval Timecode (VITC). Still other digital recorders record the timecode into the video meta data. The SYNC HD treats all timecodes as LTC or VITC. Unless you are using a VITC recorder, the timecode input will be LTC.

Timecode from the Pro Tools system comes from the HD software in conjunction with the SYNC HD.

Either system can be locked to the other. If the video is the master, pressing play on the video recorder will cause the Pro Tools system to roll forward.

Conversely, if the video is locked to Pro Tools, pressing play on Pro Tools will cause the video recorder to roll forward.

There are several options for how you can use this function. First, the video recorder can be remotely controlled from the Pro Tools transport controls. This is great for cueing up a tape. To remotely control the video recorder, select "machine" from the Transport Master submenu in the Pro Tools Transport window. Now, the Transport window controls the recorder but not Pro Tools. Make sure the remote switch on the recorder is in remote. All of the buttons should now control the video recorder. Be careful with the record button. It will cause the video recorder to insert audio or video or even go into record depending on what is armed on the recorder.

Locking picture to Pro Tools

To lock the video to Pro Tools the most common interlock is to use the transport controls to control Pro Tools and lock Pro Tools to the timecode coming from the video recorder.

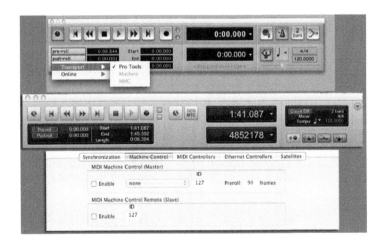

Figure 1.9 Pro Tools 7 (top) and 8 (center) Transport window. The Transport Master can be set to Pro Tools or Machine. When the transport is controlling video, the Pro Tools system is locked to the video timecode. In either case, the transport control seems to control both systems.

Set the Transport Master to Pro Tools. Now click the Online icon on the transport (Small clock, command-shift-j). The clock will flash blue and there should be a message in the edit window, "waiting for sync." When the play button is pressed both Pro Tools and the recorder will go into play. It will seem like the transport started Pro Tools but this is not the case. The transport started the recorder and Pro Tools chased after the recorder. When clicked the play button flashes until the video and Pro Tools lock into sync.

In this mode, the record function is set to Pro Tools so that you can record new tracks. In fact, everything will be as it normally is except that the video recorder is locked to Pro Tools.

It is also possible to control the recorder directly and let Pro Tools chase the timecode from the recorder. This has the advantage of now controlling the record functions on the recorder rather than Pro Tools. Simply set the Transport Master back to Machine as we did when we were using the transport controls as a remote for the video recorder. But this time we will keep the Online icon selected in the Transport window. Now we have remote control of the recorder with record controls, but Pro Tools will still follow the recorder and lock to it.

The tracks can be armed from the Machine Track Arming window. Depending on what machine is interlocked, this will show all of the available tracks, even video. On video recorders, there is also an Assemble mode that puts the recorder into record, wiping out the video, audio, and even the timecode track. It is extremely unlikely that you will ever want to record video or use the Assemble mode from Pro Tools, so be careful. Also, when working with a new setup, be careful – if the incorrect Machine Type is selected in the Machine Control page of the Peripherals window, you may be arming the video track or entering Assemble mode when you think you are arming something else. Look at the video recorder and make sure the proper tracks are armed before making an insert.

In the Transport window, it is possible to set in points and out points, and the recorder will punch in and lift out seamlessly at those points. Or you can punch in manually with the record button.

It is possible to interlock any machine with a 9-pin remote, even audio recorders like DA88. With the proper Machine Type setup in the Machine Control page of the Peripherals window, all eight audio tracks will show up in the Machine Track Arm window.

With the new Satellite Link option in Pro Tools 8.0, it is possible to lock any LE system to a second computer used only as a QuickTime video player, or up to five HD systems together. For more complex setups, many devices can be locked to Pro Tools and Pro Tools can be locked to many devices using the SYNC HD. For example:

- A film projector or magnetic film recorder (dubber) with servo can be locked to Pro Tools by feeding the biphase clock from the SYNC HD to the projector or dubber and setting the film device to servo lock. Any number of such devices can be included in the servo serial chain. Any one of these devices can be the master controller by taking it out of servo lock and setting the Pro Tools Transport Master to Machine in the Transport window.

- A digital recorder capable of resolving play and record can be interfaced with the SYNC HD. For example, a DTRS 8-track recorder like DA 98, DA78, or DA88 with a resolver board can be either the master or a controlled device. Just like a video recorder, it needs to be connected and locked to house sync, send its LTC to the SYNC HD, and be connected and control-led by the 9-pin remote.

- Several Pro Tools systems can be locked together by linking their SYNC HDs. In the event you need truly massive track count and/or multiperson mixing setup with many monitors, this is an option. On very large dubs, often one Pro Tools system is interlocked into a chain of several systems to function simply as the insert recorder in Destructive Record mode. This results in all of the comp tracks and stems being on that system's drive or drive array, ready for delivery and archiving.

It is also possible to interlock devices without using the Pro Tools interlock. Depending on the device, it may be possible to put it into a chase mode directly. For example, a DTRS with timecode interlock like DA98 can chase any timecode. Simply run a BNC cable from the Pro Tools timecode out to the DA98 timecode in. Lock the DA98 to house sync, put it into chase mode, and it will be interlocked as a controlled device to Pro Tools. A timecode offset can be entered from the DA98 menu if needed. (The menu tree of the DA98 is reminiscent of the hedge maze in *The Shining*, but it can be done.)

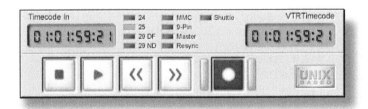

Figure 1.10 The Virtual VTR control panel.

For video machine control, there are several hybrid systems out there as well. When using the SYNC HD, it is possible to lock several Pro Tools systems together, one of which can be used as the video playback. In this type of setup, CPU usage may be no issue. It is also possible to use the SYNC HD and lock to a separate CPU using Virtual video tape recorder (VTR) (Mac only). The second CPU becomes a dedicated QuickTime video recorder interlocked with machine control.

In Pro Tools 8, this feature is available without the use of the SYNC HD: the new Satellite Link option can lock up to five HD systems together and Pro Tools 8 LE can lock to a second dedicated video system.

It is also possible to play the video directly from a disc recorder. One of the best of these systems is the Doremi V1. This is a videodisc recorder with 9-pin control. When using a disc recorder as the video playback in Machine Control,

Figure 1.11 The Doremi
V1 Video Disc Player
"video server."

it behaves exactly like a QuickTime movie: you have instant lock and the picture is always in sync with Pro Tools. It does require recording the video to the disc, but then the video is ready to go online. The drives are removable and can stay with the project. Disc recorders and virtual VTRs are the systems of choice at many high-end facilities.

Using alternate sync references

House sync is most often used on interconnected devices; however, several other sync references are available and need to be used in certain situations. Often it is not necessary to have positional reference, but it is necessary to have speed reference. For example, when transferring sync audio before it has been synced to picture, there is no positional reference as yet but it is necessary to keep it at the proper speed. Several things can be used as the speed reference on systems equipped with a SYNC HD:

- The computer processor's clock
- House sync (or any other black burst signal)
- Pilot reference from line power, pilot recorder reference (or pilot derived from blackburst)
- Word clock or digital audio input

LE systems can only lock to the computer processor's clock or a digital audio input. At different times, we may need to lock the speed of Pro Tools to an alternate clock. This is often the case when transferring the sync production audio into Pro Tools, also referred to as transferring "dailies."

1.4 SYNC HD connections on synchronized HD systems

Even if you are using an existing system and not doing any hookup, it helps understand how to use the software if you understand the cabling and hookup. The SYNC HD connects to the computer with a special Pro Tools cable from the DigiSerial port on the SYNC HD to the Core Pro Tools card installed in the computer.

When using Machine Control to interlock a video recorder or other controllable device, a 9-pin cable must be connected between the SYNC HD and the video recorder.

It is also necessary to provide timecode reference from the video recorder to the SYNC HD. On recorders that have VITC timecode, connect any video out on the recorder to the SYNC HD video in. Because VITC timecode is embedded in the video, this serves as the timecode reference. It is necessary to select LTC or VITC in the Pro Tools Peripherals window. The video connection on the SYNC HD must be terminated, if nothing will be connected to the pass-through video out, a terminator must be plugged into the connection. On recorders with LTC, connect the video recorder's LTC out to the LTC in on the SYNC HD or USD. Pro Tools can "chase" any timecode supplied to these ports. Assuming that the device sending the timecode is locked to the same speed reference, any time the Online icon in the Transport window is selected (small clock), Pro Tools will wait for timecode at this port and then follow it. If the incoming timecode is from a device that is not locked to the same speed reference, sync will drop and the system will stop after a short time as the two systems drift apart.

It is also necessary to set up a "sync loop," a simple BNC video cable from the SYNC HD Loop Sync Out connector to Loop Sync In on the Pro Tools Audio Interface and another from the Pro Tools Audio Interface Loop Sync Out to the SYNC HD Loop Sync In. Every Pro Tools I/O device must be included in this loop. This enables you to set any device in the loop as the Master device. For example, if you want to use the ADAT digital audio input on the number 2 Audio Interface as the clock source, you can set this interface as the Loop Master. The entire loop including the SYNC HD now locks to this audio input.

There also needs to be a speed reference into the SYNC HD. In a large facility, this will be the house sync, whereas for a small installation, this can be a simple black burst generator. Be sure to match the sync reference to the video format (NTSC or PAL).

1.5 Configuring the Peripherals window

For machine control, click the Machine Control tab to display the Machine Control page. Check the check box for 9-pin serial (9-pin Machine Control). The SYNC HD has two 9-pin ports making it possible to have two different video recorders or other devices connected for machine control. Select the port for the desired video recorder or other device, 1 or 2.

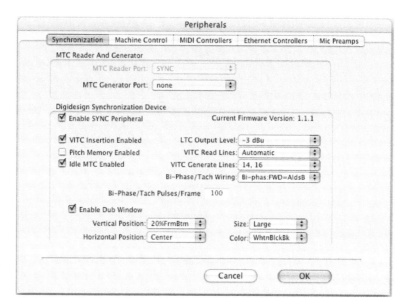

Figure 1.12 You configure the SYNC HD and Machine Control from the Peripherals window. On HD systems, in Setup go to Peripherals and click the Synchronization tab. Choose Enable SYNC peripheral (to select SYNC HD as the synchronization device on the DigiSerial port).

Now you need to set the type of recorder you are using. As most recorders use the Sony control system, even if you have another type of recorder it will probably work if it is set to Sony 9-pin. Try several Machine Type settings; BVW–75 is usually good on most recorders. Again, this is a Sony standard. You may need to set a preroll of 30 frames or so. This can shorten the time needed for the video and audio to lock together when the system rolls forward. There is always a second or so before the picture and sound lock together.

It is also possible to configure the Remote section of the Peripherals window so that Pro Tools can be controlled by the video recorder or any 9-pin device. This is rare, but you may need to control Pro Tools from another system or control console, even from another Pro Tools system. This can be configured in the 9-pin Remote section of the Peripherals window.

1.6 Configuring the Session Setup window

Depending on the needs of the session, you may need to change several settings in the Session Setup window. On many sessions, the default settings will be fine. First let's check the timecode offset and frame rate.

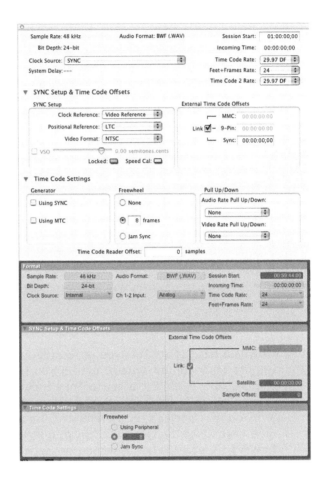

Figure 1.13 Session Setup window HD. The Session Setup window is found in the Setup menu in V7 and the Window menu in V6. From here you can setup many of the parameters of the session.

Session start

Normally, the session start is set to 01:00:00:00 with the "Picture Start" frame of the Society of Motion Pictures and Television Engineers (SMPTE) count-down leader also at this timecode on the videotape. If this is not the position of the Picture Start on the videotape, an offset needs to be set into the Pro Tools session. In the upper right you will find the Session Start field; this will likely show 01:00:00:00. Timecode from the tape is shown directly below that. If an offset needs to be set, park the tape on the Picture Start frame and highlight the Session Start field. To copy the timecode setting from the incoming timecode into the Session Start field, press = on the keyboard fol-lowed by Return to confirm the setting. Because this copies the incoming timecode from the tape into the Session Start window, the timecodes will now match. There will be a dialogue asking if you want to Maintain Timecode or Maintain Relative Position. Any audio already in the session will need to move to a new position to keep its relative position to Picture Start and take on a

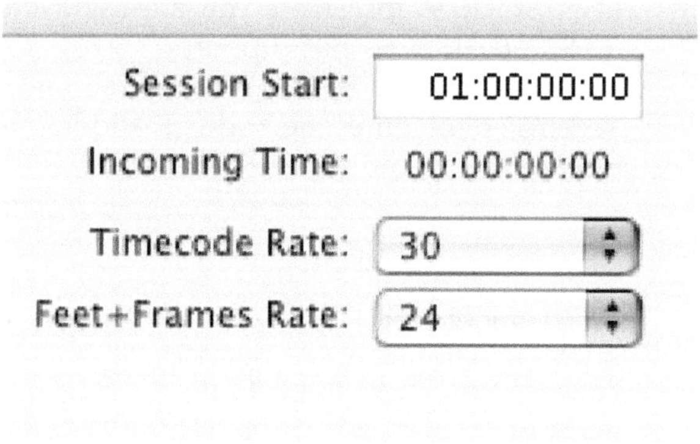

Session Start:	01:00:00:00
Incoming Time:	00:00:00:00
Timecode Rate:	30
Feet+Frames Rate:	24

Figure 1.14 Session Start TC when using Machine Control. If the timecodes match between the videotape and the Pro Tools session, then there is no need to change the session start or offset.

new timecode. Do not select Maintain Timecode as this will keep all audio regions at their current timecode position and move them out of sync.

You may not want to change the session start timecode. 01:00:00:00 is the expected norm. It is also possible to offset the two timecode values numerically. In the External Time Code Offsets section, you enter an offset timecode value in the 9-pin setting. Offsets are also settable for MIDI Machine Control (MMC) (MIDI) and Sync devices like SYNC HD or USD. If the link box is checked, any setting here will adjust all three settings. The offset is entered as the offset timecode. For example, if the video tape is 10 min 12 s and four frames ahead of the Pro Tools session, enter -00:10:12:04. This may require some fancy math, but timecode calculators are available as applications and Mac dashboard widgets.

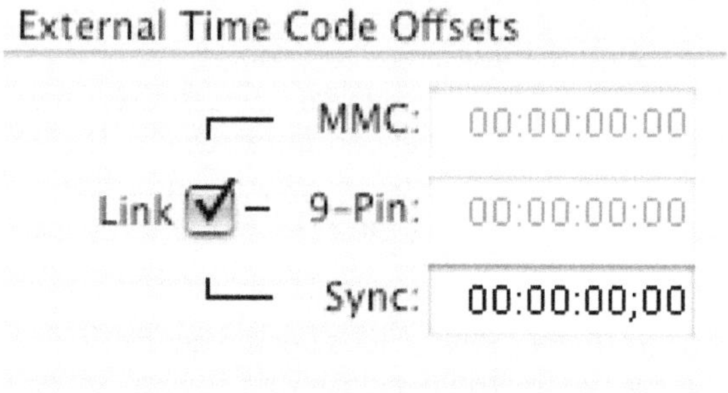

External Time Code Offsets

	MMC:	00:00:00:00
Link ☑ —	9-Pin:	00:00:00:00
	Sync:	00:00:00;00

Figure 1.15 It is also possible to offset the two timecode values numerically. In the Session Start Offset section, you enter an offset timecode value in the 9-pin setting.

The timecode rate or Frame Rate setting is used to set the frame rate on SYNC HD systems and the timecode rate on LE systems. Frame rates are not adjustable on nonsynchronized systems. Only a sync device like SYNC HD can adjust the frame rate. It is possible to display a different timecode format that will cause the counters to move at a different speed and create timecode numbers to match the desired frame rate. For example, if you are working on a 24P project (23.976 fps), your video and Pro Tools will need to be playing at NTSC 29.97 fps unless you are using the SYNC HD, which can support 23.976 (USD does not support 23.976). You may want to set the counters to 23.98 nondrop frame (NDF) to see the actual timecode positions. On LE systems, some timecode and feet-and-frames settings require the DV Toolkit option.

The clock source and Sync Setup

When using a SYNC HD or USD (V6), on HD or time domain multiplexing (TDM) systems, this will normally be set to SYNC HD (or USD). This places control of the "sync loop" in the Sync Setup selector here in the Session Setup window.

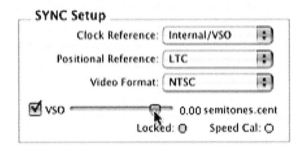

Figure 1.16 Sync Setup in the Session Setup window.

The other possible settings for the clock source are Internal and External. Both of these are setups on LE systems as well. Internal refers to the internal processor clock source. This is the usual setting on LE systems and locks Pro Tools to the computer clock. The external settings are various digital audio inputs on the audio interface (S/PDIF, AES/EBU, Tascam Digital Interface [TDIF], and ADAT). If a digital connection is made to one of these inputs, sync will need to lock to the external digital audio source.

With the SYNC HD selected as the clock source, the Clock Reference can be set in the Sync Setup. Several settings are possible. The SYNC HD or USD (V6) can be locked to video (house sync), internal/variable speed override (VSO), or any clock source in the sync loop. The positional reference can be set to any timecode input (LTC or VITC) for machine control positional reference. The video format of the machine control video is also set here. The VSO is also set here. This sets a limit on the amount of speed variance that will be applied before overriding the resolver. When you select a valid sync source, the Locked and Speed Cal light will be lit solid.

Timecode settings

The timecode generator can be used to generate LTC and MIDI Timecode (MTC) from the SYNC HD. Using Sync outputs, timecode from the device is selected in the Synchronization section of the Peripherals window. MTC to port outputs MIDI timecode to the MIDI device selected in the pop-up menu.

Freewheel sets the time Pro Tools will continue to play if it loses timecode. If there is a drop in timecode from the external timecode source (e.g., the video playback recorder), Pro Tools will stop and display an error message unless the freewheel is set to allow a number of frames to play in spite of the loss of timecode. The default setting is eight frames and the maximum is 120 frames.

Pull-up and Pull-down

The Pull-up and Pull-down settings here in the Session Setup window change the playback sample rate. This causes the session to play faster or slower. Depending on the project's workflow, this can be used to edit a film speed project at video speeds or vice versa. It is critical to understand the project's workflow to determine when and if these settings should be used. For a complete explanation of pull-up and pull-down, see Chapter 2 "Pull-up and pull-down workflow." Normally, this only changes the audio playback speed; however, there is also a setting for video pull-up and pull-down. Changing the video speed is an exotic workflow and is not advised.

1.7 MIDI interface

To use the MIDI controllers or record MIDI tracks, it is necessary to have a MIDI interface. The Mbox 2 and 003 interfaces include a MIDI interface; older Mboxes and HD systems do not come with a MIDI interface. The Digidesign MIDI I/O provides 10 MIDI inputs and 10 outputs. It connects via USB and will work with any Pro Tools system.

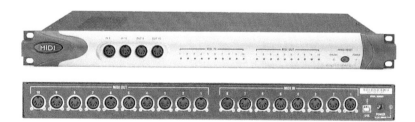

Figure 1.17 The Digidesign MIDI I/O.

While the Digidesign MIDI I/O is the gold standard, any MIDI interface will work to interface the Pro Tools system to MIDI. Digidesign also makes the M-Audio UNO, a compact, single-port, 16 input and output MIDI interface for less than $50.

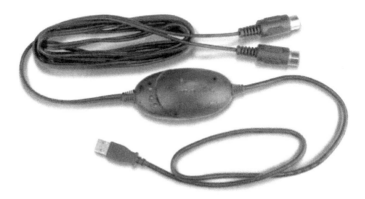

Figure 1.18 M-Audio UNO MIDI I/O.

MIDI is a serial system: 16 devices can be controlled from a single port by "daisy chaining" them from the MIDI inputs to the MIDI outputs. The MIDI devices in the chain may include sample players, keyboards, recorder controls, automated mixers, and control surfaces. MIDI can do many of the control and interface functions of the SYNC HD; however, it does not resolve speeds. This means that MIDI can be used to control only, all devices must be locked to the same speed reference and run at native speed. It can be used with any Pro Tools system, LE, M-Powered, or HD.

With a MIDI interface installed, Pro Tools can be interlocked to a MIDI-controlled device. MIDI control is set up in the Peripherals window.

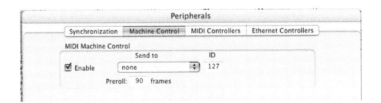

Figure 1.19 MIDI Machine Control (MMC) window in Peripherals.

MIDI Machine Control is used on some recorders and other remotely controlled devices. This is no different than working with the 9-pin remote, only the protocol is MIDI, and so there is no need to set up the control protocol. Every MIDI device in a MIDI chain needs to be assigned an ID number. The MIDI number and device name are set on the installation of the MIDI device driver. Simply enter the name of the device from the pull-down menu; the ID number of the device will appear, and it is connected.

Figure 1.20 In version 7, the MIDI control is set from the Transport window just like the Machine Control. Simply set the Transport Master to MIDI. These are set in the Peripherals window in Pro Tools 8.

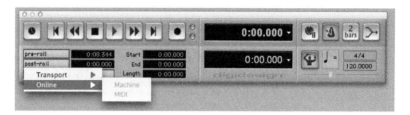

Pro Tools, MMC, and Machine Control can be used at the same time by setting MIDI and/or Machine as online in the Transport window.

MIDI also uses a timecode format, MTC. Just as many devices can "chase" Pro Tools by chasing the timecode from the SYNC HD, many devices can chase MTC. In the Session Setup window, set generate MTC to any available MIDI port and connect the device via MIDI cable. Set that device into MTC chase and it will follow Pro Tools.

Pro Tools can also chase MTC. This can provide a way to lock an LE system to an HD system with a SYNC HD. Set the HD system to generate MTC in the Peripherals and Transport window. In version 7 LE, set the Clock Reference in the LE Transport window to Generic MTC reader. Interlock by setting the LE transport control to Online (small clock). In version 8, MIDI input is set in the Peripherals window, Machine Control settings. MIDI master or slave functions can be enabled here and the drop–down menu from the V7 Transport window can be found in the master area. Both systems will need to have a MIDI interface and be connected by a MIDI cable.

Figure 1.21 MIDI controllers. MIDI-controlled devices are configured from the Peripherals window, MIDI controllers area. From here, MIDI-controlled devices such as MIDI control surfaces can be added to the MIDI control.

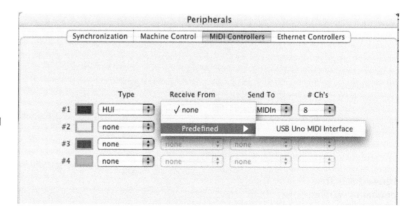

It is also possible to use MIDI as a 9-pin machine control on video recorders and other devices that use 9-pin remote control. A software application called Gallery 009 converts MMC to 9-pin control at a computer serial port. (This may require a USB to serial adaptor.)

Ethernet control

Some control surfaces, such as the C|24, connect by Ethernet rather than MIDI. These are configured from the Ethernet Controllers in the Peripherals window. The configuration is for the most part the same as MIDI, but the connection and control are via Ethernet. On networked systems, this may require the installation of an Ethernet hub; however, if the system is a stand-alone CPU without any networking, and if only one Ethernet device is connected, the connection is made with a simple cat 5 cable.

Figure 1.22 The Digidesign C|24 control surface uses Ethernet to communicate with Pro Tools.

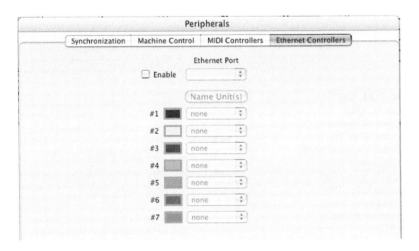

Figure 1.23 The Peripherals window. Ethernet setup area.

1.8 Data storage

Figure 1.24 Network storage used for data storage with imbedded backup can serve several Pro Tools systems.

In many large facilities, the digital audio and video is stored on a large array connected via fiber channel to all of the editing systems. It can serve the data to all of the systems at the same time and self-recover in the event of a drive crash. This is great system, however, expensive and often unnecessary. On smaller home systems, the session may be stored on the main hard drive or even on a removable FireWire drive. Pro Tools sessions can be edited from most FireWire drives, however, some brands use a FireWire bridge that is not compatible with Pro Tools.

Even on smaller Pro Tools systems, multiple drives will improve performance and can provide some level of backup. When a drive crashes (notice I don't say if), a backup of everything can avoid an hour of simulated mental illness. It is possible to configure Pro Tools to record to several drives and store backups of everything to second or third drives. In large facilities, several systems can be connected to network storage via fiber channel. On single user systems, multiple drives can connect to the computer via SCSI or even FireWire 800. There is no need for fiber channel on a single editor system.

1.9 I/O Setup

The I/O Setup window found in the Setup menu allows the user to create many different I/O setups. The I/O Setup window tells Pro Tools which audio channels are routed to which channels on the audio interface. Depending on the type of audio interface and Pro Tools version (M-Powered, LE, or HD), there may be two, eight, or vastly more inputs and outputs on the audio interface or interfaces.

Even on a simple two-channel Mbox, it is possible to have many channels controlled through the I/O Setup, but it is only possible to input or output two at a time.

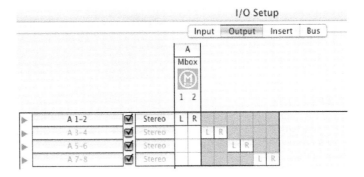

Figure 1.25 Standard I/O output settings on an Mbox with LE software.

Figure 1.25 shows the default Mbox output I/O using LE software. Notice that there are eight channels of I/O even though there are only two stereo channels on the Mbox. In this case, the L and R of the Mbox are set to outputs one and two from Pro Tools. The other I/O paths are grayed out and unavailable. These would be available if the LE audio I/O were a 003. It is possible, however, to drag the L and R from path two, set to Pro Tools channels three and four, to the left and assign them to L and R on the Mbox. Pro Tools will not let you close the window, and therefore set this as the current I/O because this will create a conflict, two Pro Tools output channels set to both the R and L of the Mbox. But if the blue check box is un-checked on the first path, this will disable that path, allowing Pro Tools channels three and four to be set to R and L on the Mbox.

The advantage of this system is that a multichannel project can be edited on a stereo system. In fact, any Pro Tools setup can be edited on any system within limits. Let's look at a 5.1 mix on an Mbox when the new 5.1 option for Pro Tools 8LE is not available.

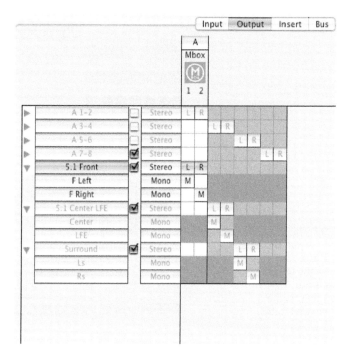

Figure 1.26 Even though LE software does not support 5.1 surround sound, it is possible to create a 5.1 I/O that can be opened later on an HD system. In this example, we are monitoring the front right and left.

In Figure 1.26, the original output I/O channels for outputs one through six have been made inactive. Channels seven and eight will not be used and have been left alone. A new path has been created using the New Path button. Two new subpaths have been created below that. The stereo path has been labeled 5.1 Front and the mono subpaths have been labeled F Left and F Right.

Two more paths have been created each with two mono subpaths. These are labeled Center, low-frequency effects (LFE), Ls, and Rs. Their assigns have been dragged to output channels three, four, five, and six. All of these channels will now appear in the mixer output assigns. Because we are editing on an Mbox, we will only hear two channels in playback, in this case the front right and left. Any tracks assigned to the center, LFE, or surrounds will be muted as these paths are inactive. If the session is opened on an HD system as a 5.1 session, these tracks can be made active and will play from the proper speakers. We will look more closely at this in Chapter 8 on 5.1 mixing.

It is possible to monitor these inactive tracks by setting the corresponding outputs to the L and R of the Mbox in the I/O settings. This will require making the front right and left inactive as only one output can be assigned to each channel in the I/O settings.

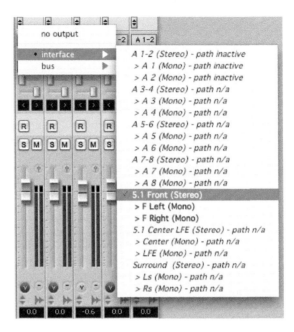

Figure 1.27 The output of any channel can be routed to the proper output from the output pull-down menu. The menu now reflects the added I/O output paths.

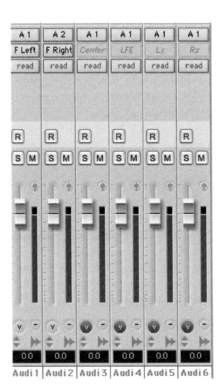

Figure 1.28 In this example, six tracks have been assigned to all six outputs on the 5.1 I/O paths. Notice that the tracks assigned to outputs not available for monitoring on the Mbox are grayed out and will not sound.

Figure 1.29 In this example, the 5.1 front pair has been made inactive by deselecting their blue check box. This makes it possible to drag the left and right stereo assigns for the center and LFE to the left and monitor these from the Mbox. It is only possible to monitor two output channels at a time on the Mbox. Other LE hardware interfaces can monitor more than two channels at a time; however, only LE systems with Pro Tools 8 and the new Complete Production Toolkit for Pro Tools 8 support 5.1 mixing.

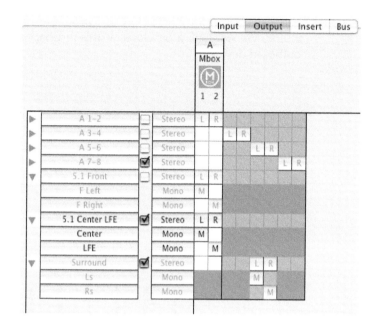

It is also possible to set up custom input I/O configurations or rename the inputs to fit your studio. For example, if you have certain hardware more or less permanently connected to your Pro Tools audio interface, you can change the names of these inputs to indicate what devices are connected to which inputs.

Figure 1.30 On this eight channel I/O, the inputs have been named to indicate what devices are connected to the Pro Tools audio interface. When the input assignments are used on the Pro Tools mixer, they enunciate what devices are available on which inputs.

The inserts I/O Setup makes it possible to use any open input–output pairs as an insert channel for inserting a hardware device into any Pro Tools mix channel. Simply make both the input and output on the available channel inactive and then set that pair as an insert in the Insert tab of the I/O Setup window. Now any hardware device, for example, a spring reverb, connected between that input and output is available in the I/O section of the mixer inserts. We will look a bit closer at this in Chapter 8, in the section on using outboard devices.

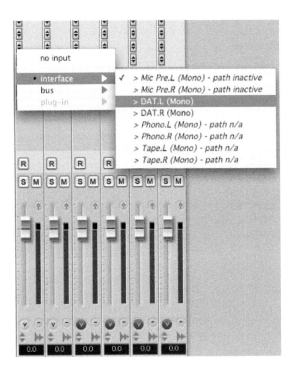

Figure 1.31 The names set in the I/O Setup now show as named inputs on the mixer and edit window input and output settings.

1.10 Studio design

There was a time when studio design was not an issue while editing sound. Most motion picture sound editing was done with headphones, and only one track could be heard at a time. Once edited, the entire project was mixed in a huge dub theater where all tracks could be interlocked and played in a critical listening environment. This was the first time anyone had heard all of the sounds together, much less at full fidelity. It was common for sounds to be rejected at the first attempt at a rough mix.

Over time, this workflow has evolved to the point where it is often difficult to tell the difference between the editing room and the mix theater. In fact, often they are now the same room. Large projects still mix in large dubbing theaters; however, television and smaller feature films normally mix in a "sweetening" studio, a small studio that combines the editing and mixing into one room.

Critical audio decisions are made at every phase of the edit and so the edit room must be a critical listening environment. Assuming that you have a professional sound postproduction room, then this information is probably of little use to you – unless last week your professional sound room was the store room for janitor supplies, but someone stuck a Pro Tools system in there and stuck your name on the door.

If you are editing and mixing in a spare room or garage, then you, too, may need some help. Most home studios are totally incapable of producing accurate sound: the monitor speakers are poor, the room is noisy and reverberant, or the neighbors call the police every time you try to mix 24-bit 5.1 at Dolby-specified levels.

Studio design is a complex science and well beyond the scope of this book, however, a few tips can help at least improve the situation. First, if you are planning on doing any real work on the room, get help. A consultation from a studio design firm may cost as little as $400. Second, if you can't do anything to the room – say, you live in an apartment – get help. You can edit with micro speakers or headphones, but you need to mix in a real room with real monitoring. Fortunately, with Pro Tools this is easy: you can take your project almost anywhere. If you can't afford a high-end facility, find someone who will help you out in their down time for a good price.

Basic acoustics

Standing waves, as well as waves that cancel and amplify, create problems in our mix room. A mix that sounds great in such a room will sound totally different in any other room. While we have no control over the listener's room, we do have control over the room we choose to mix in. And the more neutral our room is, the better our mixes will sound in any room. The goal in making a room neutral is to make sure no frequencies are boosted or attenuated. Most rooms are either "boomy," building some of the low frequencies, or "shrill," building some high frequencies. It's easy to test the room after you have built it to see how it turned out, but challenging to test the room before you build it.

These boosts are caused by reflections off the walls. Sound waves move exactly like water waves, only in three dimensions and much faster.

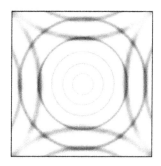

Figure 1.32 Reflected waves coming off of square walls.

Think of a square water tank about two feet on a side and with totally vertical walls. Drop a marble in the center of the tank and ripples move out from there, hit the sides and reflect back. The reflections collide with incoming waves, canceling some and making some bigger.

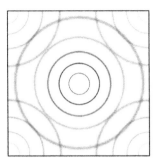

Figure 1.33 Corner loading. When waves bounce off of walls at a corner, they reflect symmetrically off the two walls and the floor creating a new set of waves that match the original.

When the waves collide with the 90° corners, they reflect out in curves just like the original sound source. These also interact with the original waves and the boosted reflections off of the flat walls. 90° corners are diabolical – they act like sonic mirrors reflecting back one-eighth of the wave shape, but mirroring the entire wave. In the above image, notice that the waves are only one-quarter of the original wave. Because waves are three-dimensional, where the walls intersect the floor at 90°, the same thing is happening. Therefore, only one-eighth of the wave is reflecting back, but the shape of the wave is unaltered, and to some extent refocused.

You can prove it with a simple experiment. Turn a speaker into a corner about a foot from the walls. Keep it flat on the floor. The highs and mids will be muted, but some frequencies of the bass will get louder. This is why some subwoofers are turned into a corner. It improves their efficiency.

Figure 1.34 Directional sound waves.

Fortunately, these problems are somewhat more contained in a real room. In the water tank, the waves radiate in all directions when something is dropped in the tank. As we can see in Fig. 1.34, speakers direct the sound in one direction. Even the bass, which tends to radiate out in all directions, is rather focused by the monitor speaker. So, in a stereo room most of the reflection problems are focused on the back walls and corners of the room.

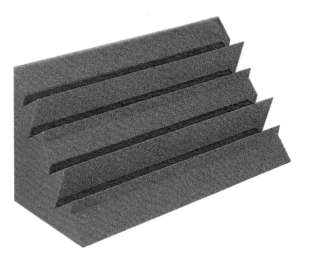

Figure 1.35 The Auralex LENRD Bass trap keeps waves from radiating out from the corners.

Reducing low frequencies takes lining the walls with absorbing material over a foot thick. Fortunately, it is not necessary to line all the walls with this thick insulation, only the problem areas. 90° corners can have a short-angle wall across them or they can have a "trap" placed in them. One version of the trap is made of foam. These need to be large – they may come out a foot from the corner but they don't need to be solid; in fact, they work better if they are more open with "fins" extending from the corner. These are available from several suppliers.

It's also possible to build a trap. Line the walls in the corner with thick sound-absorbing material. This should extend out about two feet on each wall. Then, place a framework with fabric attached across the corner.

There is another type of "bass trap." In this case, it is a deep box for deadening a wall. This will usually be the back wall, the one with the monitor speakers and the sub pointed toward it. If the waves are stopped here, the other walls will not be a big problem. Some of these look like a big pillow and work well. Some people place the pillow in a box with a fabric cover. It is usually only necessary to place a bass trap near the top of the wall extending down about halfway. Furniture and your own body will help reduce waves below that.

Several companies offer studio kits, all the traps, diffusers, and glue needed to treat a certain size room. These are handy and very cost-efficient. These can be augmented for larger rooms and modified as needed.

Figure 1.36 The RPG Studio in a Box kit.

Figure 1.37 Standing waves.

Now let's look at standing waves. If we vibrate the water tank, at certain frequencies standing waves appear. These waves appear to not move because they are bouncing off the sides at the exact speed of the vibration. Sound waves in the studio behave in a very similar way.

We have several tools at our disposal that can help. Standing waves are caused by waves bouncing off walls and bouncing between parallel walls. So, we can start by trying to reduce parallel walls. Think of stealth airplanes. They are designed to scatter waves, in this case radar waves. Some have strange angles deflecting the waves in almost random directions, others have outward curved surfaces scattering waves like a lens. Either idea can work in our studio.

Figure 1.38 Angling the walls or ceiling in short sections.

It is not always possible to totally eliminate parallel walls, but they should be avoided. Because sound is three-dimensional, the same holds true for the ceiling and the floor. In a large studio, this is not a problem; ceilings can be high and easily angled. In a garage, it's a bit tricky. Fortunately, it is not necessary to angle the entire wall or ceiling; it can be angled in short segments.

Another good tool is the room proportions. Certain proportions create standing waves at a given frequency in two or even three dimensions. Others can only cerate a standing wave at a given frequency in one dimension, and this greatly reduces the effect of any standing waves. A room that measures 8 × 11′3″ × 15′3″ or a proportion of 1:1.40:1.90 will help fight standing waves. There are other good proportions, but the real key is to avoid bad proportions. 1:1:1 is perhaps the worst proportions. Anything that is exact multiples will propagate waves at the same frequency. The 1:1 water tank with parallel sides propagates waves.

Speaker placement

No matter what you do, there will be bad effects in some parts of the room. But we don't need the entire space to sound wonderful, only the critical listening area. Everything in the room is directed to the mix chair, the placement

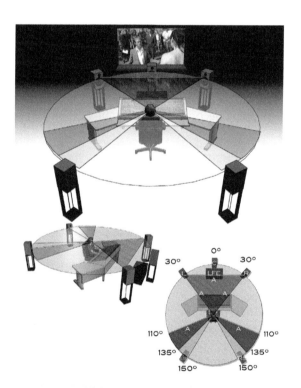

Figure 1.39 On 5.1 surround systems, the monitor speakers must be equidistant from the mix chair. Moreover, the front right and left must be that same distance apart for best performance. The rears should be not less than 110° off axis, and not more than 150°, preferably between 135° and 150°.

of the video monitor, the desk, the controls on the desk, and certainly the monitor speakers.

Stereo and 5.1 monitor placements are similar. The right and left monitors should be equally distant from each other and the mix chair. This forms an equilateral triangle. They should be above the desk and at the same height, or at least the tweeters should be the same height on all speakers. This may require that some monitor speakers be mounted upside-down.

On 5.1 systems, the center monitor should be halfway between the stereo speakers and at the exact same height and distance from the mix chair. This will place it slightly further back from the right and left monitors forming an arc. Continuing this arc forms a circle, the rear monitors should be placed on this circle. They should be at the same height as the front monitors and behind the mix chair between 150° and 110° off axis from the center monitor, preferably between 135° and 150°. The position of the subwoofer will depend on the room and the kind of subwoofer but it is less critical.

1.11 Dolby Pro Logic 5.1 playback

While the project is still in discrete audio channels in Pro Tools, 5.1 monitoring in Pro Tools HD is simple. But before the 5.1 sound can be used in a film print or DVD, the discrete channels need to be encoded into Dolby AC3 files. Until recently, this required encoding with Dolby hardware or Dolby licensed compression software such as Apple's Compressor. Once encoded, the resulting AC3 files cannot be played back in Pro Tools; they require playback through a Dolby decoder that sends the proper signals to the proper speakers.

The latest generation of Macintosh computers and some PCs have optical outputs that can send Dolby Pro Logic to a Pro Logic decoder. Although there are many Pro Logic preamps and decoders on the market at very reasonable prices, Pro Logic playback in Pro Tools is more complicated than simply attaching such a device to the optical out on the computer.

The problem is the Dolby AC3 files are exported as QuickTime files, and QuickTime does not support Dolby 5.1. It therefore sends no signal to the optical out. If the AC3 files are burned into a DVD, any DVD player with Dolby out can send the 5.1 to the decoder or preamp. Apple's DVD player does support Dolby AC3 files and does send the 5.1 out through the optical jack on the computer. This also means that if you connect a Dolby decoder to the optical out on the computer as well as to your monitors, you can watch 5.1 movies from DVD in your mix room, which may be a negotiating point when you want to build your mix room in the basement or garage.

There are now plug-ins for encoding Dolby Pro Logic directly in Pro Tools, and these plug-ins can play back the encoded files. True, they can't play the latest 20-hour cut of *Lord of the Rings* from Blue Ray, but when it comes to professional audio encoding, being able to encode and check the playback right in Pro Tools is a vast improvement over the older workflow. We will look at these plug-ins in Chapter 4, Pro Tools Plug-ins.

1.12 Isolation

We also need a quiet place to work. We can't have any outside noises or machine noises. For some reason, people generally think that if you want to make a wall "sound proof," all you need to do is hang some insulation on it. Granted, this helps, at least a bit. But think of it this way, if the neighbor is having a party and you don't want to hear it, are you going to close the window or the drapes? While there are reasons to deaden some of the walls with absorbing material, when trying to stop sound from going through a wall, soft and porous aren't going to do much.

Unfortunately, sound can travel through almost anything. Even if you had six inches of concrete in the wall, some low-frequency sound would still get through. When the neighbor's kid drives by in his "awesome ride" with hip–hop cranked to eleven through his fifteens at 1 kW, from a block away, you hear his bass – not the mids and highs – the bass. Low frequency punches through almost anything and travels far. Ultra highs can be stopped dead by a piece of cloth.

There are special sound treatments that you can put inside of a wall. Because of the popularity of home theaters, such products are widely available. Some of these products resemble plain plastic sheeting, yet these air-tight membranes do wonders for mid-low frequency isolation. Other products are heavy foam composites in several layers each with different sound transmission qualities for different frequencies. The wall should be as thick as you can make it, 6" minimum. Because sound can travel through wall studs, two walls with a space in-between is better than a single wall, or the wall studs can be "staggered," that is to say, one forming the inside of the wall, the next the outside, inside, and so forth. This is not as good as two walls, but it is better than a single wall.

It may also be necessary to "float" the walls. If any sound can get into the floor, then it will travel up into the walls and radiate out from there. This is also true in the other direction. Unless the floor is a solid concrete pad directly on the ground, the walls need to be floated. This is simple; there are many products available for this. These are rubber pads or channels that go between the floor and the walls.

The equipment also needs to be isolated. Even a cooling fan is more noise than you should have in the room. Ideally, the computer and any noisy equipment should be in an isolation room or "machine room." It is also possible to use an isolation rack, an equipment rack with a glass door, and isolation material in the case. You also need to isolate the air-conditioning and heating. Special baffles are available for this.

Ergonomics and equipment

It is important to be able to reach all of the gear that you use all of the time. You also need to be able to see the video as well as the computer screens – and it seems that there are never enough computer screens. Just keeping the edit and mix windows open pretty much ties up two screens. While many people get by with two, or even tough it out with one, three allows you to have everything at your fingertips.

A control surface greatly improves the access to the controls and speeds the mixing process. This is especially true on film mixes. While music mixes may be set for long sections, on film mixes everything is going up and down all over the place all the time. Trying to set up an adequate mix with just a mouse is a study in frustration and a waste of time. The control surface allows you to adjust several things at the same time, hear how they interact and make exact changes.

Many people have several pieces of hardware "gadgets" that they like to use. While these may be available as plug-ins, the sound of the analog device or specific digital processor may be only available by having and using the stand-alone device. These devices will also need to be patched. The controls and patches of these pieces of equipment need to be where you can get to them from the mix chair. The "usual" location has been either right in the mixing console or directly behind the mixing chair in a short rack. While at one time tens of such devices were found in the mixing theater, today most of this function has been taken over by plug-ins and only a few devices remain. Needless to say, all of these devices need to be totally quiet or else mounted in an isolation rack.

Many studios will also have a MIDI keyboard. Even if this is not used for music, it may be used to add Foley sounds or special effects. Here, too, this needs to be where the mixer/editor can reach it from the mix chair.

Audio monitors

It is assumed that audio for film and television will be played on full-range speakers, not satellite speakers with a subwoofer. In the theater, this is always the case, but in the home this is rarely the case. This should not affect the mix; however, the satellite systems manage their own crossover, routing the bass from the stereo pair to the sub at the frequency needed for that particular system. This is also true for surround systems and mixes. Where we run into a problem is when we are surround mixing on a satellite system. The patching from Pro Tools requires that the six outputs should be sent to the six speakers. The LFE channel is not a satellite subwoofer, it is only used for audio effects. In such a setup, the only audio sent to the LFE is from the LFE channel, which does not contain any of the bass of the stereo pair or any of the surrounds. None of the bass from the stereo pair that would normally be routed to the satellite sub is routed to the LFE unless a crossover is running inside of Pro Tools or as an external hardware system to do just that. In the mix, if we simply route some of the bass from the other five channels to the LFE to "fill out" the sound, when this mix is played on a full range system or a satellite system, the bass is double what it should be. What we need is a satellite crossover system for our mix room that will either operate inside Pro Tools or actively on the six-channel speaker output.

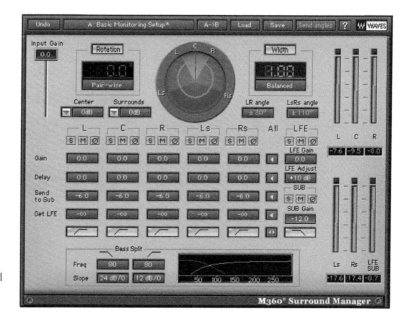

Figure 1.40 The Waves 360° sound manager can adjust timing if the speakers are not at their optimum position as well as manage the LFE as a satellite subwoofer.

The Waves 360° Surround Tools Manager plug-in splits the LFE and functions into a satellite sub and LFE at the same time. When we mix sound to the LFE only that sound is recorded into that channel. But also a controllable amount of bass from the other channels is sent to the LFE speaker so that it also functions as a satellite sub. The Waves 360° Surround Tools plug-ins set also comes with many other great surround mixing tools.

Tuning the audio monitors

Once the room is finished and the audio monitors are in place, it is possible to test the room for problems. Pink noise is played through the system and analyzed with an audio spectrum analyzer. This shows where there are spikes and valleys in the spectrum. If the acoustics has been properly designed, the room should be fairly neutral. If there are massive problems, this may require rethinking the traps. But if the problems are small, they can be compensated for with environmental equalization. This is done with a graphic equalizer installed in a place where it will not be "messed with." (Some people can't resist sliding a slider.) Most of the tuning will be in the low end, which is good. Equalizing the highs messes with the stereo imaging. As always with equalization, less is more. The better the room, the better the sound.

Figure 1.41 Radio shack sound pressure level (SPL) meter.

The level of the audio monitors also needs to be set. There needs to be a zero marking on the audio monitors' level control. All of the audio monitors should be heard at the same level at the mix chair. One exception may be the LFE. If no management system is being used like Waves 360° Surround Tools Management plug-in, the LFE may need to be set at +10dB. Some LFE-amplified speakers have a +10dB switch that can be set. Bass needs this +10 boost to be perceived as the same volume as the other monitors, as bass is perceived at a lower level by the human ear. When in doubt, check with the manufacturer. The SMPTE has developed standardized monitoring levels for cinema postproduction work. For film mixing, the reference level is 85 dB SPL at the mix chair. This is read with a simple sound pressure level meter (SPL) set on the C weighted/Slow scale. An affordable model is available at Radio Shack. The reference level for television is 80–85 dB and 85–95 dB for music production.

Set each monitor individually by playing pink noise from the Pro Tools AudioSuite signal generator at the default −20 dB. Shut down all other monitors other than the one being calibrated at the power amp. Now set the volume of this monitor to the desired level (85 dB for film). Repeat this for the remaining five monitors. If the subwoofer has a +10 setting, set this level with the switch on. While this sounds wacko, we are using the C weighted scale that takes this roll-off into account. If you add the +10 after setting the level, you will get too much sub. In fact, no matter how you set the sub, it will not be all that accurate; you really need to calibrate the room with the spectrum analyzer, but this is more or less close depending on the room.

Once the individual monitors are set, set the main volume to the proper level (80–85 dB for television, 85 dB for film, and 85–95 dB for music production) at the mix chair.

Video monitors

The choice of video monitor is a very subjective decision. In a postproduction facility, this will normally be a reference video monitor; however, because no video coloring or any video processing happens here, the monitor can be just about anything that shows detail and doesn't make any noise. The second consideration is a large one – we have gone to a lot of trouble to make the room as quiet as we can, we don't want to bring in something that has a fan in it. Also, as the monitor will have many hours put on it, we may not want something that costs a lot to operate.

The obvious choice is a CRT. These can be anything from a reference monitor to a large screen HD television. The down side is that they are deep, heavy and are only available in sizes less than 40 inches. But they are cheap to operate, have a sharp picture, last for years, and are very quiet.

Plasma monitors come in sizes up to 70 inches, they are thin and can be hung on the wall. They are sharp, but have a rather short life and in some installations make noise. At very high elevations, they whistle and are impossible to even be near yet alone try to make critical audio decisions around. But they can make a good monitor in the right situation.

LCD flat panels have all of the advantages of the plasma screens without the down sides. They are now available in the giant sizes plasmas offer, look great, and are becoming quite affordable.

Video projectors are able to project any size, look great, and take up very little room. They can be very expensive to operate; some bulbs cost over $1000 and have a life of only 1000 hours. The biggest problem with projectors is their cooling fans. They cannot be in the mix room unless they are mounted in an isolation "box," and this must be well-ventilated or the unit will over-heat. The best solution for installing a projector is to place it in the isolation room, or in its own isolation room or "booth," and this complicates the room design. Projectors have a limited throw to certain screen sizes, so the projector must be matched to the installation.

Figure 1.42 Juniper Post Studios Mix Theater.

A large mix theater may have several mix positions at the control surface and two or three computer monitors for each position, all displaying several Pro Tools systems connected together. Huge films often require huge resources. But whether big or small, the mix room needs to be efficient and comfortable.

The garage studio

If you are editing and mixing in the garage, you have a little money to spend on the project and your family is OK with parking outside for the next 20 years, joy! Welcome to the fraternity of Garage Audio Studios (GAS), which is the smell you will need to get out of there before beginning any work. There are several other things it would be nice to get out: the washer and dryer, the extra refrigerator or freezer, the water heater, and the lawn mower.

After negotiations with your family, you will settle for the lawn mower and bicycles going into a shed. Actually, this is good, because you have given in on some things you wanted, so you can cash that in later when you get the credit card bills.

Depending on your state and city, you may have a nonstarter. In some places, you will never be given the permit, and construction without a permit may result in tearing out everything you have done and could affect your insurance. Find out what the law in your area is so that you are well informed before building anything. Also think safety. Even if you have been given a permit, this does not mean that your design is safe. There are no windows in a studio and often only one door. Don't sleep in there. Have fire extinguishers. Have a way out.

You need a quiet space to be able to hear exactly what you are doing, and the freezer, washing machine, and water heater don't seem conducive to that. But then neither does a computer cooling fan, videotape recorders, other cooling fans, or the heating/air-conditioning.

What you need is a machine room: a place to put all of the noisy equipment. Since the washer and dryer are already in place, the location of the machine room may be dictated to you. You also need to keep the outside sounds outside and the inside sounds inside.

Fortunately, because you are in a garage, you can build a box inside of a box. Before building the box, you will want to do everything you can with the garage door and any windows. In this case, foam insulation may be about all you can do short of walling up the door. What ever you do, try to make it airtight and solid. Even a small gap at the bottom of 1/4" will equate to 54 square inches, a hole big enough to stick your head through.

Once you have done everything, you can, with the basic garage, build the studio space inside of it with 2×6 stud walls.

Because you are building directly on the cement floor, it's not likely that any significant noise will transfer from the floor up into your walls. If you are building directly on a wood floor or prestressed concrete with space below, any sounds in the floor will transmit directly up into the walls and into the room. In this case, it is important to "float" the walls on isolation material. Several products are available.

The studio "box" will mostly fill the garage space with some space left for the machine room/laundry/storage area. The "air space" around the box can help in isolation, especially if the new wall has an exposed absorbent material. Any sound entering the space will bounce back and be partly absorbed.

The door into the studio is critical; the average door will let almost all of the outside sounds in. However, if the washing machine is elsewhere, and all you need to worry about is the computer and recorder sounds, a good outside-grade door with a good seal may do the job. There are "sound-proof" doors available, but they are expensive.

The shape of the box should not necessarily reflect the shape of the garage. Curved walls are wonderful if they curve away from the room and not concave inside it. They are tricky to build but possible. Avoid parallel walls and 90° corners wherever you can. And, again, a little help from a consultant will really help come up with a good room.

Depending on the kind of sound quality you want in the room, you will be deadening some surfaces. It's unlikely that you will want a totally dead room. When placing absorbent material on the walls, the highs and mids are killed more than the bass, and this creates a "muffled" sound. Normally, areas of the wall are left hard to bounce highs and mids. Other special diffusers are used on the walls to help scatter the bass without killing the highs and mids. And remember, this is not a recording studio per se, it is a motion picture and television editing and mixing room. This is not to say that you will not be recording in here, but that is not its primary function. You want a very neutral listening theater. In fact, if you are having trouble negotiating for space, point out that this is really just an awesome home theater with a mixing console.

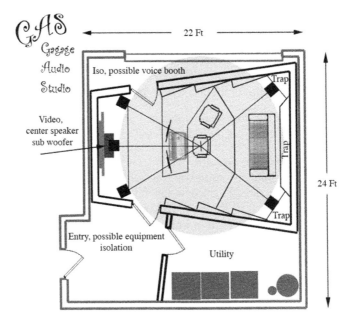

Figure 1.43 Welcome to the fraternity of Garage Audio Studios (GAS).

In this garage studio, the washer, dryer, freezer, and water heater are still located in the attached garage. The box within a box design isolates the studio from the outside, and while it is less isolated from the utility area, these can be controlled. No washing during a final mix! The wall and ceiling treatments can be any number of products, even a "studio kit" set of diffusers and traps. The isolation materials in the inner walls can also be any number of products, but they should be solid and airtight.

For larger spaces, the walls can be moved out as long as the front wall, mix position, and the monitor speakers keep this relationship. If the back wall is moved back until the mix chair is 38% of the distance between front and back walls, this is great. If the room is bigger than that, the circle and monitor placements should be made larger as well.

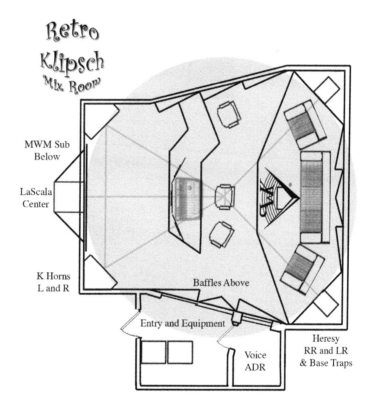

Figure 1.44 Retro Klipsch mix room.

In Figure 1.44, we have a retro Klipsch mix room. This is a bit too large to fit in the average garage, as the larger Klipsch horn type speakers require more room. The heart of this mix room is the pair of "Klipschorns," also known as the K horn used as the stereo pair. These were invented in 1945 by Colonel Paul W. Klipsch, and they are still manufactured on a limited basis. While it may seem delusional to use an antique speaker as the main stereo pair, the proof is in the sound. Klipsch speakers have a loyal following, and, yes, they (we) have been accused of fanaticism. The Klipschorn is still considered by many to be the finest loudspeaker ever made and is the only speaker in the world that has been in continuous production for over 60 years.

Figure 1.45 Paul W. Klipsch and his Klipschorn speaker (left).

Figure 1.46 Paul W. Klipsch and the tin shed where the first K horns were manufactured.

The Klipschorn was first manufactured in a tin shed in Hope, Arkansas, in 1946. "The Colonel" was a colonel in an Arkansas militia and well known as an "eccentric." An avid jogger, he would use parking meters on the main street as hurdles.

He was also known for his colorful use of language and his willingness to disagree with people. He took notes in church so as to use quotes in his rebuttal of the sermon. The corporate motto was "Bullshit" – this in response to claims made by other speaker designers. The motto was emblazoned on bright yellow buttons that were passed out when the need arose. He also had this motto stitched on the lining of his suit coat: bull on his right side, the remainder on the left. This way he could opine by merely "flashing" when the need arose.

Figure 1.47 Lascala 300.

The center speaker is a Klipsch Lascala. The Lascala was first built in 1963 for Arkansas gubernatorial candidate Winthrop Rockefeller to use as a public address system. While the Klipschorn is "corner loaded," requiring that it be placed tightly into a corner of the room, the Lascala is a free-standing version of the Klipschorn that can be placed anywhere. Hopefully, Winthrop took along some Elvis vinyl to rock the voters.

Figure 1.48 The 1957 Heresy.

The surrounds are Heresy speakers, originally designed in 1957 as a center channel speaker for the Klipschorn. The Klipschorn was originally used in single speaker mono systems, but as stereo became available the speakers were more often used as stereo pairs. Many people felt the stereo perspective lacked any real center, and center channel speakers became popular with audiophiles. They fell into disuse as phasing accuracy in recordings and amplifiers reduced the center "hole" in stereo recordings. Today with 5.1 surround, the center channel is back in a big way. The Heresy was always a little small for a center between K horns, but they are ideal for surrounds.

Figure 1.49 The MCM Grand theater system. The cabinet on the bottom of the stack is the MWM Dual.

The sub is the massive Klipsch MWM. The MWM Dual is still made; however, this single would need to be custom built. The MWM Dual is part of the Klipsch MCM Grand industrial theater system, used in high-end THX theaters around the world. The MWM Dual is 35 inches tall, 72 inches wide, 48 inches deep, and weighs in at over 300 pounds. The single is half that height.

It could be placed on the floor below the Lascala raising the tweeter of the Lascala to the same height as the K horns. It could also be mounted tight to the ceiling above the center channel.

1.13 Systems used on "Loves Devotion Forever"

Several systems were used to edit "Loves Devotion Forever." The project was stored, and principally edited from, a LaCie d2 500 Gig FireWire drive. This was backed up on two other systems. Three Mbox systems with 7.0 LE, upgraded to 7.3 during the edit, were used for transfer and some of the editing. DV toolkit allowed for timecode display and provided DigiTranslator. Two MIX 24 systems with USD and 6.4 software were used to edit, record automatic dialogue replacement (ADR), and temp mix. These were upgraded to HD 7.3 with SYNC HD, house sync, and MIDI I/O. Video was from QuickTime during edit, machine control during the final dub and Foley.

The final dub was performed at Todd AO studios, and the Foley was recorded at Film Leaders both on HD systems. Other hardware included:

- Control 24 control surface
- DA 78 DTRS recorders
- Nagra 4.2
- Sound Devices 744T
- PD 4 DAT

Figure 1.50 Brooks Institute Pro Tools 5.1 HD mix – edit room.

While some might say that setting up the gear is half the work, others will say it is half the fun. The room and equipment are an evolving project. Every year brings new options, challenges, projects, and ideas. Several years ago, one good film recorder cost as much as the expanded Pro Tools rack in Fig. 1.6. And because of the affordability and flexibility of digital gear, even the smallest system is a work in progress.

In this chapter

2

Audio workflow

Film and video workflow has become a complex issue as film and video editing has moved into the computerized nonlinear editing environment. What was once a straightforward path is now a complex maze with twists, turns, and unfortunately, dead ends. Both the audio and picture face scores of possible directions, choices that can save time and money while enhancing the project, or cost money, lose time and diminish the final product.

Within the audio workflow, there are many options; fortunately, Pro Tools deals well with these options. Often, Pro Tools provides several options of its own, and while this complicates the decision making even more, it allows for the individual to find a system that works for both them and the project.

2.1 Pull-up and pull-down workflows when shooting on film

We will be looking at how to pull-up and pull-down later in this chapter; for now, we need to understand when it is necessary to pull-up and down, and by how much. When the project has been shot on motion picture film, at times it becomes necessary to change the speed at which the audio plays. We always want the sound and picture to run at the exact same speed, but at times, we need to change the speed of the picture, and this requires also to change the speed of the sound playback.

There are three basic frame rates for picture. First, the film speed of 24 frames per second (fps). Second, the PAL video speed of 25 fps. PAL is the video format used in most of the world. Film for video is also shot at 25 fps in most PAL countries. Third, the NTSC video speed of 29.97 fps. This is the video system used in North America and Japan.

When we transfer 24 fps film to a PAL video system, we speed up by 1 fps. If we copy PAL video to 35 mm film and project it on a 24 fps projector, it is slowed down by 1 fps. Because this change is fairly small, it is not noticeable in the image. But if the audio is played at its original speed, it is hopelessly out of sync in a few seconds. It becomes necessary to pull-up or pull-down the speed of the audio and do it in such a way that the audio does not sound strange. The best workflow here would be to pull the 25 fps video down to 24 fps before recording, editing, and mixing the tracks.

This becomes more complicated with NTSC video. The change from 24 fps to 29.97 fps is so large that it would noticeably alter the movements on screen if it were simply sped up. So, when 24 fps film is transferred to NTSC video, it goes through the *3:2 pull-down* process.

The film is loaded on a NTSC "telecine" machine. The device copies the film to video in fields, which are half-frames of video. Two of these half-frames are added into the video in every four film frames. This causes four frames of film to become five frames of video, which is to say, 24 frames of film to become 30 frames of video. If this video were played at 30 fps, it would actually be playing at its original speed. One second of film would still be one second of video. But NTSC video plays at 29.97 fps, not 30 fps. In order for the telecine machine to create 29.97 fps video, the film is run at 23.976 fps. This slows the speed by 0.1%. This process is referred to as 3:2 pull-down. The 3:2 expression refers to the fact that the film frames are being transferred to the videotape with a cadence of three fields from one film frame, two from the next, three from the next, and so on. Pull-down refers to the fact that the speed has been slowed by 0.1% as well. Unfortunately, the terminology has become polluted and many people now refer to pull-down as the entire film-to-tape transfer. Fortunately, this is not true in audio; pull-down always refers to slowing down the speed to match a new picture speed. Pull-up refers to speeding up the audio to match a different picture speed.

There is also a fairly new video speed known as 24P. 24P usually plays at 23.976 fps. This way the 3-2 cadence of added fields can be applied in camera and 29.97 NTSC recorded to tape. This video can be captured into the Avid or Final Cut Pro at either 29.97 or 23.976. Because it is shot in camera at 23.976, there is never a need to pull-up or pull-down.

These are all critical numbers when dealing with video and sound for film and video, but they can all be broken down into the three basic frame rates: Film, NTSC, and PAL.

For purposes of this discussion,

- Film speed will refer to any media playing at 24 fps or 30 fps.
- NTSC speed will refer to any media playing at 29.97 fps or 23.976 fps.
- PAL will refer to any media playing at 25 fps.

The most common pull-ups and pull-downs are as follows:

- Audio for film transferred to NTSC will need to be pulled down by 0.1%.

- Audio for NTSC transferred to film must be pulled up by .1001%.

- Audio for film transferred to PAL must be pulled up by 1 fps. (4.166%).

Fortunately, there are many ways to perform these pull-ups and pull-downs. Unfortunately, there are so many ways the workflow becomes a problem. If you are not careful, you might pull the audio down twice or not at all.

One system available on the SYNC HD is the ability to simply play the session at a different speed. In the Session Setup window, there are settings for these frame rates. The session can simply be set to 29.97 and the audio, but not the video, is slowed down. But this does not change the speed of the actual audio media, it only changes the playback speed. Any audio exported to another software would still play at its original speed. There are workflows where changing the playback speed makes sense, but for most workflows taking the audio to a pulled down (or pulled up) speed and editing it at that speed that makes the most sense. This is even true on film projects that will be edited at video speed and then pulled back up to film speed for film printing.

Film to NTSC pull-down

Here are several common workflows for capturing and syncing audio on a film shoot that will be telecined to NTSC and edited on Avid or Final Cut Pro:

The first workflow is the most common: syncing in telecine. This would be appropriate for finishing on film or NTSC.

- Transfer all original production audio to DTRS (DA-88) at 30 fps. (This will be done by the telecine facility.)

- Interlock the DTRS to the telecine at 29.97 fps. Pull-down is achieved in transfer or telecine depending on the telecine workflow and the original sample rate of the production audio.

- Telecine the footage and sync to the slates, either on the clapper sticks or the timecode window on the slate.

- Edit on the Avid or Final Cut Pro at 29.97 or 23.976 fps.

- Export the edited project to Pro Tools with no pull-down and perform the sound edit and mix at 29.97 or 23.976 fps. Machine control video require transferring to videotape at 29.97.

- If the project is finishing on video, this will be the finished speed.

- If the project is going back to film, pull-up the composite mix in the transfer to optical sound or DTS encoding.

The second workflow requires syncing on the Avid or in Final Cut Pro. This would be also appropriate for finishing on film or NTSC.

- Telecine all footage to 29.97 using Final Cut Pro and Avid compliance.

- Transfer all audio to Pro Tools.

- Depending on its format, the audio can be pulled down during transfer.

- Divide all audio clips from master transfers.

- Export all audio clips to Avid or Final Cut Pro.

- If the audio was not pulled down during transfer, it should be pulled down on export.

- Sync all takes to slates in Final Cut Pro or Avid using the Merge clip function (FCP) or Auto Sync (Avid).

- Edit on the Avid or Final Cut Pro at 29.97 or 23.976 fps.

- Export the edited project to Pro Tools with no pull-down and perform the sound edit and mix at 29.97 or 23.976 fps. Machine control video require transferring to videotape at 29.97.

- If the project is finishing on video, this will be the finished speed.

- If the project is going back to film, pull-up the composite mix in the transfer to optical sound or DTS encoding.

Different audio formats can be pulled down in transfer or on export. Any digital audio sampled in production at 48 048 Hz is automatically pulled down when it is transferred into a 48 000 Hz (48 k) session. This is the preferred workflow by many production sound mixers. Because this pull-down happens automatically on import, care needs to be taken to avoid not pulling this audio down twice, once in transfer and again in export. It is critical to know what sample rate was used in production.

Nagra audio can be pulled down either in transfer or on export. If the Nagra is locked to video or house sync, it will pull-down on transfer. If it is locked to 60 Hz pilot or if the SYNC HD is locked to the Nagra, it will still need to be pulled down in transfer. The techniques used for transfer of dailies with or without pull-down are covered later in this chapter.

The third workflow is a 24 fps workflow that does not require pull-down. This would only be appropriate on projects to be finished to film.

- Telecine all footage to 29.97 using Final Cut Pro and Avid compliance.

- Create a 24 fps Avid project and capture all the telecined footage. This will pull all the video speed footage backup to film speed at 24 fps. On Final

Cut Pro, in the reverse telecine process in Cinema Tools, reverse to 24 fps. On both Avid and FCP, the footage is now back to its original 24 fps.

- Transfer all audio to Pro Tools without pull-down.

- Sync all takes to slates in Final Cut Pro or Avid using the Merge clip function (FCP) or Auto Sync (Avid).

- Edit on the Avid or Final Cut Pro at 24 fps.

- Export the edited project to Pro Tools with no pull-down and perform the sound edit and mix at 24 fps. This requires playing video from QuickTime; machine control will not play at 24 fps.

- Transfer the final mix to optical sound without pull-up.

The fourth workflow is a 24 fps workflow that uses session pull-down. This would also only be appropriate on projects to be finished to film.

- Prepare and telecine the film just as on the 24 fps described previously.

- Export the edited sound project to Pro Tools without pull-up or pull-down and perform the sound edit and mix at 24 fps.

- Export the 24 fps video to 29.97 fps QuickTime or videotape for machine control. When imported into Pro Tools, this new video speed will cause the project to move quickly out of sync.

- Set the session to pull-down in the Session Setup window. This will play the 24 fps audio at 29.97 fps in sync with the picture. Video can be from machine control or QuickTime.

- Because the project is going back to film, all audio will be recorded and imported at 24 fps; it is only edited at 29.97. As this requires no pull-up of the final mix, the finished audio will never undergo sample rate conversion and be as pristine as possible.

Film to PAL pull-down

When editing film for finishing on film, PAL is not often used. Film for film finish must be edited on the Avid or Final Cut Pro at 24 fps or 23.976 fps to ensure an accurate cut list. Telecine to NTSC with reverse telecine is the standard workflow for achieving 24 fps or 23.976 fps media.

When film is shot for PAL finishing, it is normally shot at 25 fps and then telecined at 1:1 requiring no pull-down. However, when film shot at 24 fps is telecined to PAL, it is telecined at 25 fps, pulling the speed up by 4%. While this is not very noticeable in the image, when the audio is pulled up by 4% is pitch shifted up and is very noticeable. Usually, this is repaired by digitally pitch shifting back down.

Because of these problems, a new system has been developed that avoids audio pull-up entirely. Borrowing from the NTSC field cadence system, PAL is often telecined with a field cadence of 2:2:2:2:2:2:2:2:2:2:3. This creates a new field in every 12 or one new frame in every 24 converting it to 25. It is critical to know which telecine system was used on the footage to know if you need to pull-up 4% or not pull-up at all. This would only be proper for finishing on PAL video, not film.

2.2 Production recording

While this is not a book on production recording and mixing, it is important to understand the formats and timecode conventions used in production mixing to properly deal with the production media. The field recorder may be one of the following several formats:

- 1/4" analog mono reel-to-reel with pilot.

- 1/4" analog stereo reel-to-reel with pilot and timecode.

- 48 kHz sample rate DAT with or without timecode.

- 48.048 kHz sample rate DAT with 30 or 24 fps timecode.

- Data CD or DVD at 48k, 48.048k, or HD sample rates up to 192 kHz with or without timecode.

Timecode on the production audio can be drop frame (DF) or nondrop frame (NDF). Rarely, it is contiguous; normally, it has breaks between takes. The usual timecode system in production is to let the timecode generator free run, either from a preset timecode or as time of day. Because the timecode generator continues between takes, there are breaks between takes.

The timecode slate, if it is used, also has a free running timecode generator in it that is "jammed" to the generator in the recorder. A cable is plugged between the units, and the jamb button on the slate is pressed matching the number on the slate to the timecode on the production recorder. These will match sync for several hours but require being rejammed several times per day. The slate is filmed at the head or tail of every take recording in the picture what audio timecode matches the picture. Usually, the syncing is still done to the clapper sticks on the slate, but it can be done to the numbers on the slate. The slate does serve to inform the sync editor of the timecode at sticks close so that this number can be entered into the database and therefore the audio EDL.

Time of day needs to be time accurate to match the time of day clock, either NDF at film speed 24 fps or 30 fps, or DF at video speed 29.97 fps or 23.976 fps. Only these timecode formats are time accurate. If, however, these formats are pulled down, they become a new format. The video speed formats become nonexistent formats, 29.6703 DF and 23.73624 DF. Therefore, these timecode formats should not be used with pull-down. The film speed formats of 30 NDF and 24 NDF become video speed formats 29.97 NDF and 23.976 NDF. This requires using free-run timecode.

When using a free-run timecode generator, the format does not need to be time accurate, so any format can be used. Here, too, if video speed formats are pulled down, they become nonexistent formats that should not be used. There is also the strange 30 DF format that is a nonreal format until it is pulled down and it becomes 29.97 DF.

On some recorders, it is also possible to "oversample" by 0.1% so that the audio is automatically pulled down when played back at the normal sample rate. Typically, this is 48 048 Hz to be played back at 48 000 Hz.

This table shows combinations of sample rates and timecode formats that can be used in production with and without pull-down.

	Time of day or free run	Pulls down to	Free run	Pulls down to	Sample rate (Hz)
No pull-down (video)	29.97 DF	–	29.97 NDF	–	48 000 or
	23.98 DF	–	23.98 NDF	–	96 000
24	24 NDF		24 NDF		48 000
					96 000
Pull-down (film)	30 DF	29.97 DF	30 NDF	29.97 NDF	48 048 or
	24 DF	23.98 DF	24 NDF	23.98 NDF	96 096
PAL	25 NDF		25 NDF		48 000

2.3 Transferring dailies with or without pull-down

Production audio may be delivered on several different formats. Some may have timecode, others will not. Most are digital; however, analog reel-to-reel is still used and even preferred by some production mixers on some projects. Pull-up and pull-down are issues when transferring dailies. It is critical to understand the workflow of the project.

Dailies may be 1–6 tracks. The tracks may be discrete from several microphones worn on the actors, they may be the overall audio from a boom mic, from microphones hidden on the set, safety tracks, or any combination.

Generally speaking, all tracks will be imported into the same number of tracks with the possible exception of any safety tracks. The safety track(s) are recorded at −10 dB from the main track and are used where the primary track is over modulated. Depending on the workflow, the safety track(s) may be transferred with the other tracks and synced with the others, or kept as a safety backup on the master production tape only. So, depending on the workflow, you may be importing from one to six audio tracks.

Managing an audio edit decision list

Depending on the workflow, it may be necessary to create and manage an audio EDL (edit decision list). The audio EDL is managed in either Avid or Cinema Tools on Final Cut Pro projects. If the audio is synced in telecine, importing the telecine logs into Avid or Cinema Tools creates the EDL. On projects where the production audio will be imported in Pro Tools and synced on the Avid or FCP, the audio timecode will need to be entered into the database by the sync editor. If a timecode slate was used in production, the audio timecode can be read from the slate at sticks close and entered into the database. If not, the slate sticks close should be logged from the production tapes.

On tapeless data recorders, the media files can contain the timecode and other information as metadata. The Broadcast Wave Format (BWF) contains a wealth of information as metadata that includes timecode. The BWF metadata can be read in Pro Tools as timestamp information. BWF metadata can also be read in Avid, but when using Final Cut Pro, the audio metadata must be converted into a self-contained QuickTime file. One software package for achieving this is BWF2XML. There are other software packages that can be used to read all the metadata in the BWF files to aid in file management.

The advantage of the EDL is that there is now a map to every frame of every take. If the dialogue editor wants to retransfer or check the original production audio, they know on which audiotape and at what timecode the audio can be found. It is often necessary to make a log of the clapper close frames as the audio is transferred, even on projects that used a timecode slate. On some projects, all the edit audio will be replaced with retransfers of the production audio; the EDL is essential to recapturing the original audio. Usually, timecode production audio is synced in telecine, avoiding the need to create a database by hand. We will look at using the audio EDL to recapture the production audio in Chapter 5 "Dialogue Editing and Replacement."

As with any database, it is only as good as the data it is managing. Everyone who logs tapes and media or labels tapes must take extreme care and be 100% accurate. Also, care must be taken when working with or transferring the media not to corrupt or modify the metadata.

Often, tape audio is transferred to DTRS or DAT tape before capturing and syncing. This tape, also called a "simuldat" is then used as the master audiotape. The simuldat is recorded at film speed with matching timecode from the original production audiotape. The playback can be locked to house sync causing the tape to pull-down to video speed while locked to Pro Tools. Or the DTRS can be played at its original speed by locking to a nonvideo speed reference. These are most often used when syncing in telecine because of ease of pull-down and the ability to lock to the telecine machine.

On the rare production audiotapes where there is contiguous timecode, it is possible to match the timecode of the HD Pro Tools media with the timecode on the production masters. This is often the case when filming concerts or other live events, where the production recorder never cuts. Patch the timecode out on the audio player into the SYNC HD timecode in and place the Pro Tool Transport Online (small clock). Enter the beginning timecode from the production master tape into the session start timecode in the Session Setup window. Pro Tools will follow the playback of the production master, and the original time stamp of the Pro Tools media will match the production master tape. When this media is cut into subclips, the time stamp will still match the original production timecode.

Analog recordings with pilot

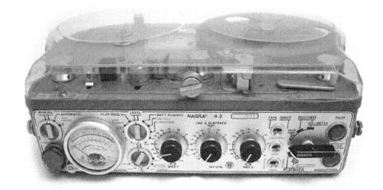

Figure 2.1 Nagra 4.2 1/4" analog recorder.

On older analog reel-to-reel tape recorders like the Nagra, sync is supplied by recording a "pilot" on the tape. This is a 60-cycle signal generated by a crystal clock in the production recorder. In playback, this is compared to a reference pilot. If the two match, the playback is in sync. If the pilot is speeding up or slowing down, the playback machine will change the playback speed until they match. This playback feature is called a resolver.

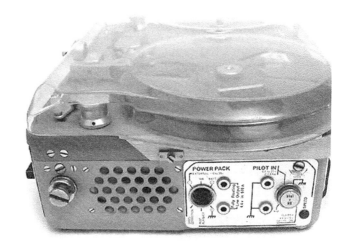

Figure 2.2 On the side of the Nagra 4 are several pilot inputs and outputs. If stripped house sync is supplied at the pilot in pin on the Xtal connector, the resolver will pull-down the audio to video speed and lock to house sync. If 60 Hz pilot from the SYNC HD is supplied, the Nagra will lock at the original speed to the SYNC HD.

The SYNC HD has an input for this pilot. If the Nagra pilot out is connected to the SYNC HD pilot in and the SYNC HD clock reference is set to this pilot, it will produce a 100% sync accurate transfer.

If you need to pull-down the Nagra audio as it is transferred to Pro Tools, locking the SYNC HD to house sync, and then using the house sync as the pilot reference for the Nagra will achieve pull-down. This requires a device that "strips" the video from house sync and creates a 59.94 Hz cycle pilot reference for the Nagra. This is connected to the pilot input on the Nagra, which will now lock to house sync and run slow by exactly 0.1%. Such a system can hold sync indefinitely.

It is also possible to "soft lock" the Nagra to Pro Tools by locking the Nagra to its internal crystal clock with the internal resolver on. Lock Pro Tools to any sync, house or internal. Now transfer with each system locked to it own reference. This will not produce a 100% sync accurate transfer, but it will hold sync + or − one frame for several minutes. As these are dailies, if the takes are not over 5 min, it will be frame accurate sync.

Some Nagras have two tracks and timecode. In this case, the timecode may need to be managed and two tracks transferred into Pro Tools. A simuldat with matching timecode will ease this process. Simuldat is managed like any other DAT production audio.

DAT production audio

DAT uses several sample rates not only as a way to record the audio but also as a speed reference. 44 100 Hz is used in CD production and is considered a "home" nonprofessional format. DAT recorders also use a 48 000 Hz sample rate that has been thought of as the "standard" sample rate for film and video production. Some DAT recorders can also use a 48 048 Hz sample rate to achieve pull-down on playback (For a more detailed explanation of digital audio, see Appendix 2 "Digital Audio").

In playback, the DAT locks to the sample rate on the tape and compares this to its internal clock in the same way a Nagra compares the pilot on the tape to a known reference. If the audio is transferred to Pro Tools through the analog audio connections and Pro Tools is locked to internal or house sync, then the speed is soft locked just as with the Nagra soft lock; and here too, this cannot hold sample accurate sync at all and can only hold frame accurate sync for several minutes.

Few DAT recorders have the ability to lock to an external speed reference, so the usual system involves locking Pro Tools to the DAT's internal clock. The digital output on most DAT recorders is Sony/Phillips Digital Interface (S/PDIF). If this audio source is used for the transfer, then the speed reference of Pro Tools can be set to this input locking Pro Tools to the internal clock in the DAT.

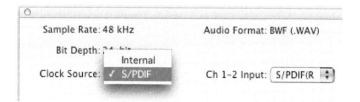

Figure 2.3 Locking the session to S/PDIF for DAT transfers.

On LE systems, this is set in the Session Setup window. On HD systems, the audio IO being used for the S/PDIF input must be selected as the sync loop master in the peripherals window and the IO input set to S/PDIF digital rather than analog.

There are several ways to record the DAT into Pro Tools. Depending on the recording of the DAT production audio, there may be index points marking the breaks between takes, beep cues, or nothing at all. The takes can be recorded into Pro Tools one at a time, or the entire tape recorded into the session and broken up into subclip takes from there.

Audio in need of pull-down not recorded at 48 048 Hz will be pulled down in export. (This is covered at the end of this section.)

Digital tapeless systems

The newest digital audio recorders record directly to flash memory or drive. In this case, the audio can be imported directly into the Avid, Final Cut Pro, or Pro Tools. Many sample rates are available, including pull-down sample rates. The Sound Devices 744T offers 32, 44.1, 48, 48.048, 88.2, 96, 96.096, 176.4,

Figure 2.4 The
Sound Devices 744T.

and 192 kHz. When the 48.048 and 96.096 are imported into a 48 or 96 K session, they play pulled down.

These systems record in BWF. This digital format sounds great and can be encoded with metadata (data about the audio clip). This can include timecode, the scene and take information, camera information or just about anything the recordist chooses to include in the slating. When this audio is imported into the latest versions of Avid and Final Cut Pro, the actual BWF audio is used, with the metadata intact. This is not a transfer per se but an import of the bit-by-bit production audio. When the audio is exported to Pro Tools, the original production audio with metadata is still being used. This more or less eliminates the need to ever go back to the production audio and retransfer, as the editor is working with an exact copy of the production audio at all times.

The timecode and other information are recorded into the BWF audio metadata and can be read by various software systems. Delivery will usually be on data DVD or CD. The audio is imported from the optical disc or other delivery system into Pro Tools using the import audio to region list command in the file menu.

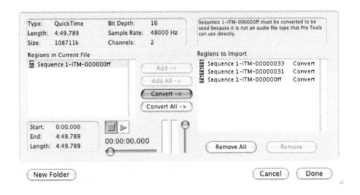

Figure 2.5 Import audio
to region list dialogue.

Pulling down in Pro Tools

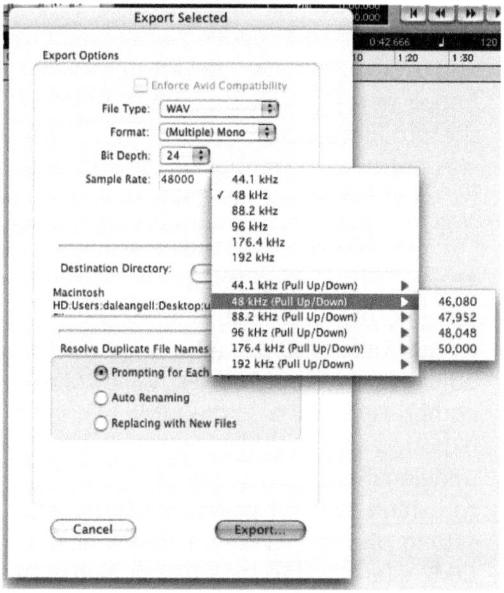

Figure 2.6 The audio can be pulled down as it is exported to Avid or Final Cut Pro in Pro Tools.

It is critical to know when to pull-down. Everyone in the postproduction loop must be aware of the workflow to avoid failing to make a necessary pull-down or inadvertently make a double pull-down.

To export with pull-down, select all the regions in the region list that are to be exported and use the Export Regions as Files command in the regions list pop-up menu.

From the Export Regions as Files dialogue, you can opt to pull-up or pull-down by 0.1% or 4%. On 48 000 Hz audio, select 47 952 to pull-down by 0.1% or 48.048 to pull-up. To pull-up 24 fps to 25 fps for PAL video, select 50 000.

Although this is an acceptable way to pull-down dailies, it is not an acceptable way to pull-up an entire project to print back to film. For final pull-up, see Chapter 9 "Output and Delivery."

2.4 Exporting the locked cut to Pro Tools

There are several ways to export the audio from the Avid or Final Cut Pro locked cut to Pro Tools. With the EDL workflow only an EDL is exported, and the production audio is recaptured from the EDL into Pro Tools. The EDL is a text file that describes the in and out points in the edit, what tape the audio came from and its original timecode on the production tape. It is often in the CMX 3600 format; however, several formats exist. This puts control of the transfers and handling of the audio into the hands of the postproduction audio crew. If there are any problems with the audio, the audio transfer people at least know that they have the original and best media. For example, the production mixer may have recorded a safety track at −10 dB, but this was not used in transfer or the edit. If the track is overmodulated in the edit, the safety may save the take. Only by returning to the production media can the sound editor be sure they have the best audio available.

This has been the standard workflow for decades but represents a large amount of work that is unnecessary when using an OMF (Open Media Framework) or OMFI (Open Media Framework Interchange) file. When the edited project is exported as an OMF, all the audio and video media are compressed into a single OMF file. Often, only the audio used in the edit is exported and "handles" are added to every audio region. Handles are simply additional audio media at the head and tail of the region making it possible to extend the edit point as a function of the sound edit. The default handle setting on both Avid and Final Cut Pro is 1 s. Some edit systems can export OMF referenced to the original media. In this case, the OMF contains only the edit information and not the actual media. Most OMF files are self-contained and do contain all the media as well as the edit information.

Although OMF compresses some types of audio, the process is lossless. One potential problem with OMF is that the audio from the syncing and editing is used in the final mix. Any problem with this audio caused by errors in transfer or handling of the audio will also find its way into the final mix. Therefore, a combination of the two systems may be best. The OMF can be imported into Pro Tools and the EDL used as a guide to the original audio to allow retransfer of any problem audio.

When using BWF production audio with OMF, the BWF is simply encapsulated in the OMF, keeping the BWF production audio intact. The OMF is used as the session audio files with the OMF simply linking all timeline regions to the audio in the OMF. When using the latest versions of Final Cut Pro and Avid, the BWF audio can be imported and used in the edit without transcoding or altering it. The OMF still contains this BWF audio, and when it is used as the audio in the Pro Tools session, you are in fact using the original production audio.

It is always necessary to have a countdown leader edited into the project. The editor should have placed the "Picture Start" frame of the leader as the first frame of the project and the leader with the same frame rate as the edit. If this is a DV project or DV down-converted project, you can use a QuickTime of the edit as the Pro Tools video reference. Non-FireWire formats such as Digi-Beta must use machine control.

On DV projects, export the video and audio as a QuickTime movie onto a FireWire drive. Pro Tools will be able to play your film in 29.97, 23.98, or 24 fps. On projects that will use machine control, a tape dub to the format of the Pro Tools machine control must be made from the Avid or Final Cut Pro project. The "Picture Start" frame of the countdown leader should be at timecode 01:00:00:00 on the videotape.

From the Avid or Final Cut Pro, export the audio as an OMF to a portable FireWire drive. The length of audio handles is set in the export OMF dialogue. Any nested sequences from the edit will be combined into a mixed track. Most often, you want to be able to control the individual tracks and not have them premixed. The visual editor should un-nest these tracks before export. Crossfades also are exported on most systems and may be unchangeable in Pro Tools. Be sure to remove any crossfades from the edit before the OMF export.

The Avid "auto sync" function can be used to preserve the original metadata including the original timecode. Referenced media also contains the original metadata because it is referenced to the original media as opposed to converting the media into a self-contained OMF file.

Most digital audio workstations can convert and import OMF files, not just Pro Tools. It is commonly used as a portable universal transport format. Converting the OMF into a Pro Tools session is done with the DigiTranslator Pro Tools option.

The Final Cut Pro or Avid audio tracks can also be exported as Audio Interchange File Format (AIFF) files and imported into Pro Tools. This imports the tracks as continuous audio regions through the entire project, requiring that the sections containing audio be edited from these continuous regions. This is easily done with the Pro Tools Strip Silence function. A major downside of importing AIFF files is that audio is brought into Pro Tools with no handles. Also, BWF audio will be transcoded into AIFF, losing the metadata. One advantage is that the audio can be time-stamped before cutting it into regions, ensuring that all regions have a unique time stamp.

2.5 Importing the audio from the picture edit, OMF versus EDL

To convert an OMF, simply go to open and navigate to the OMF file. This will launch DigiTranslator 2.0 and create a new Pro Tools session. DigiTranslator 2.0 is optional software from Digidesign. Set your session settings and name the session. Click OK and this will open the DigiTranslator

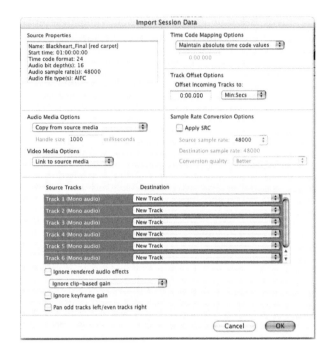

Figure 2.7 DigiTranslator is used to import OMF and AFF files into Pro Tools. DigiTranslator is part of the DV Toolkit option as well as a separately available option.

dialogue. From here, you can convert your OMF into a Pro Tools session. Do not import any scratch or guide tracks you don't want imported into the Pro Tools session, but select Import as New Track for all wanted tracks.

After the Pro Tools session opens, if you are using QuickTime video playback, import the QuickTime movie into the session. For improved performance, you may want to copy the QuickTime movie to the hard drive and import it from there. If you are using a video monitor for playback, make sure that play to FireWire is selected. The movie will now play on the NTSC monitor. Check the 2 pop on the countdown leader. As both the audio and video were the same length when exported, they should be in sync.

Figure 2.8 Import audio from movie is used to import the embedded audio from the QuickTime movie.

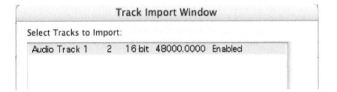

Also import the audio from current movie. This will import the audio from the QuickTime movie, all the current tracks mixed down. This will import into the region list. A new stereo track needs to be created and the region placed in the new track. Check the 2 pop for this audio and make sure it is in sync with the video and all the tracks. This will serve as a reality check: a way to check sync on any tracks you may inadvertently slip out of sync. Make the track inactive or active any time you need to compare sync to this guide track.

To capture from an EDL, the EDL is used as a guide to import or record the original production audio from the BWF files, the Nagra tape, the DAT, or the Simuldat into Pro Tools and recreate the picture edit in Pro Tools.

Different picture editing systems manage the EDL in different ways. Final Cut Pro can manage two different EDLs, the EDL in the Final Cut Pro edit and a Cinema Tools audio EDL. The Final Cut Pro EDL will not show the original timecode or audio reel information if that information was not present in the audio metadata. This requires using BWF production audio. Any timecode audio can be managed from the Cinema Tools database if that information is entered in that database. Imported FLX files from a film telecine contain this audio timecode, but on many projects, the timecode and reel information must be written into the database manually.

We will be looking at how to use the EDL to reconstruct the audio edit in Chapter 5 "Dialogue Editing."

Changes to the locked cut

While the sound is being edited, the picture edit may be changed for a variety of reasons. Even though the picture is "locked" these days, it is often thought of as "latched." Historically, making changes was a big problem for the sound editors and composers; however, technology has made the change process smoother and simpler.

The biggest problem will be for the composer. Depending on the placement of music and changes, it may require that several music cues will need to be rewritten and rerecorded. Or it could mean simply retiming the length of the session to match the new picture.

The sound editors can use special software to facilitate the changes. A system called "Virtual Katy" works with Final Cut Pro or Avid change lists and conforms the Pro Tools session to the new picture. Changes can also be made manually from the change lists. A new reference video should be exported from the picture edit to ensure it is accurate to the changed edit.

2.6 Timestamping

When audio is recorded into Pro Tools, it is "time stamped" or embedded with timecode information that records its position in the edit. When Pro Tools is in Spot mode, this address is displayed in the Spot Dialog window. If a

Figure 2.9
Timestamping audio regions allows the audio to be placed back into sync with Spot mode.

region is moved out of sync, it can be selected in Spot mode and the original time stamp recalled by clicking on the original time stamp recall.

Self-contained, OMF-derived audio regions have no "time stamp" if they are moved out of sync; the only way to get them back into sync is to compare them to the guide track. A time stamp can be added in the regions list pop-up menu with the "time stamp selected" command. This sets the current position as the user time stamp. To recall the user time stamp, again select the region in Spot mode and click on the user time stamp recall.

BWF production audio synced in Avid using the auto sync function and only referenced by the OMF will still have its original metadata timecode that will show as the original time stamp in Pro Tools. When using Final Cut Pro, BWF audio metadata can be converted into a self-contained QuickTime file using the BWF2XML application before importing into Final Cut Pro. However, the timecode is striped by the self-contained OMF export, to reacquire the timecode, the original BWF can be imported into Pro Tools using the OMF media and the EDL as a guide. This is simple with Avid edits, it may be more trouble than it is worth on Final Cut Pro. In this case, timestamping may be the best system for moving things back into sync.

You are now ready to edit your sound design. All the audio tracks from the locked cut edit and the video are now in the Pro Tools session.

The sound design will consist of three basic elements: dialogue, sound effects, and music. These three basic components have subcomponents, but the three basic elements must always be kept in their own area and eventually mixed into their own master tracks. These components, known as *stems*, are an asset of the project, and when a buyer makes a deal on the motion picture, they expect these assets to be included as separate components. A distributor wants to cut trailers and spots, and this is only possible if the stems are separate. Moreover, the film may be "dubbed" into other languages, in which case only the effects and music (the M & E) will remain. It will require extra sound work to prepare the M & E for the foreign language dub. Sound effects from the production tracks that ended up in the dialogue stem will need to be replaced. A film may be sold to a foreign market even before it is shot, so often a complete M & E is cut in the sound design; however, it may be mixed later after the mix, also known as the *dub*, is finished. The tracks containing the complete finished mix are often referred to as the "comp dub" to help distinguish it from the stems.

2.7 Portability

As the project moves forward, it is usually necessary to be able to move the project from one location to another. There may be a need to record Foley at a stage in Los Angeles and score music in the Bay Area, all while editing effects in Oregon.

Figure 2.10 DigiDelivery can be used to move audio and even video between several Pro Tools systems.

The ultimate in Pro Tools portability is achieved with DigiDelivery. DigiDelivery allows for the uploading of any audio files to the DigiDelivery server. The upload is encrypted for security. At the moment of uploading, everyone in the postproduction loop receives e-mail from the server with an encryption key that un-encrypts the file and downloads it to that Pro Tools system. If eight sound editors are working in eight cities in the United States and music is being scored in Europe, everyone has access to everyone else's audio as it becomes available. The time-stamped audio regions only need to be manually placed in the proper tracks, and now everyone has the exact same session.

This system was used effectively on the *Lord of the Rings* movie. The mix was being performed in New Zealand as the score was being recorded in London. Picture changes were even being made and reconformed with Virtual Katy, and everyone had access to the changes or new audio as it became available. A videoconference link was also used between London and New Zealand so that face-to-face communication was possible.

Digidesign sells the DigiDelivery servers; however, it is not necessary to own one to use DigiDelivery. It is possible to open an account on a third-party DigiDelivery server. The file senders need to set up a file transfer software system to that account; however, file recipients need only a fast connection and an e-mail account.

The DigiDelivery Client Software is a free download on the Digidesign.com Web site. It is available for Windows XP, Mac OS 9, and OS X.

It is also possible to link several Pro Tools systems together from any location in the world. In some setups, the Internet connection is by ISDN and is somewhat complex and expensive to use. There are per minute charges on ISDN bandwidth.

A simpler solution is Source Elements Source-Connect, a Pro Tools plug-in capable of streaming audio and even controlling one Pro Tools system from another. This only requires a fast connection (DSL, Cable, or T1). The media being streamed is a compressed Advanced Audio Coding (AAC) audio; however, the actual audio files can be transferred via DigiDelivery after the session. The advantage is being able to direct and record voice-over or automatic dialogue replacement (ADR) remotely from any Pro Tools system anywhere.

There is a tremendous amount of latency or lag in the connection, but the user is never aware of it. Both systems must contain the same session and media, then the timecode is streamed from the recording system to the control system. The record system's timecode and audio arrive together and interlock in sync with the playback of the control system. The two sessions communicate with each other via the talkback system. Transport control can be from either system, the control system can be directing the session or controlling the system. Several systems can be linked, however, only two can be controlled at the same time.

Portable systems

It is entirely possible to run an LE session on a laptop with an Mbox 2 or Mbox 2 Pro interface. This does not provide the highest performance in terms of number of tracks and plug-ins, but for many projects, this will provide all the power needed and the entire system is now portable. Moreover, the laptop can be used as an external drive to a larger system making it possible to open the LE project from the laptop on an HD system without even dumping the session to the HD system drives (although you may want to copy the project the hard drive anyway). Any part of the portable session can be imported into the HD session using the Import Session Data function.

A project or even a duplicate project on a laptop with an Mbox has the advantage of being a complete portable Pro Tools studio with 2 mic inputs, headphone monitoring, battery operation, and video reference. It is possible to record effects or even Foley with this system anywhere, assuming you have a quiet place to work. With the DV Toolkit option installed in the LE system, timecode or feet and frames can be displayed in the LE session.

It is also possible to run the entire project from a removable FireWire hard drive. If the video playback is from QuickTime, copying the QuickTime from the removable drive to the internal hard drive and reimporting it into the session will dramatically improve the performance.

If it becomes necessary to open the session on an older Pro Tools system, this can be done by saving a copy in the older version from the File menu. This may not solve all the problems when attempting to open a project on an older system, many functions will not work in the older version but at least it should open.

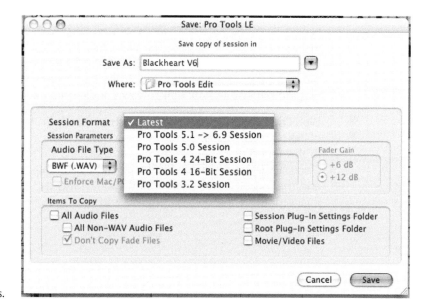

Figure 2.11 By using "save a copy" it is possible to convert to older versions of Pro Tools.

2.8 Workflow on "Loves Devotion Forever"

"Loves Devotion Forever" was shot on 35 mm film at 24 fps. Audio was recorded on PD 4 to DAT at 48 048 k sample rate. All audio was copied to DA98 DTRS and all slate claps were logged against their timecode. Negative was cut into selects and outtakes and the selects were workprinted. Workprint was projected silent at FotoKem labs. The select rolls were telecined at Laser Pacific interlocked to the DTRS audiotapes, syncing to the camera slates and the clapper logs. Audio was pulled down in the transfer to DTRS.

An error occurred during telecine that was not caught until later. Because the negative had been cut to print the selects, care needed to be taken in telecine to re-enter the keycode numbers after each splice. The telecine machine can read the barcodes that accompany the keycodes, but they are not all read by the telecine, most are interpolated from the known numbers. So if a splice is missed, it will be followed by several feet of wrong numbers. This was the case in one scene on "Loves Devotion Forever." At one point, two shots had the same keycodes, and this showed up as a double use. The fix was to pull the negative and read the keycode numbers at sticks close on the slate of the two shots and enter the correct information in the database.

The audio information was fine, but it shows that any database is only as good as the data. The same thing can happen to the audio information – if extreme care is not taken to avoid mistakes, huge amounts of time can be wasted figuring out what is wrong and repairing it.

"Loves Devotion Forever" was edited on Final Cut Pro at 23.976 fps. All data, including the audio timecodes, were managed in Cinema Tools. The picture was exported as QuickTime and printed to 29.97 Beta SP for machine control. Although the 29.97 fps and 23.976 fps frame rates did not match, the speed was the same so that all pulled-down audio matched.

Workflow is often confusing, but this comes from lack of understanding and planning. It is critical to plan out the best workflow with input from everyone in the sound edit loop, and it is necessary that everyone stays with the plan. Don't assume anything (yes, we all know that this "makes an ass out of you and me"), but it wastes time and money and can impact the quality of the final product. Have a working plan, and stick to it.

In this chapter

Editing Tools and the Pro Tools Interface

3.1 The edit window

Timecode display

Editing for film and video is normally done to timecode. This is not available on M-Powered and LE systems; however, it can be added to LE systems with the DV Toolkit 2. Timecode allows for frame-by-frame and subframe nudging, grid settings, and timecode spotting. This is not to say that film and video cannot be cut to minutes and seconds, but timecode is generally used and is standard procedure. It is therefore what everyone else on the project will be using. Film has been edited to feet and frames as well, available on HD systems and LE systems with the DV Toolkit, but this is falling into disuse.

When using the DV Toolkit, the Timecode Rate for the session can be set in the Session Setup window with the Timecode Rate selector just as on HD systems. The Timecode ruler displays the Time Scale in SMPTE frames (hours:minutes:seconds:frames). Also, with the DV Toolkit, the pull-up and pull-down features are available. These are covered in Chapter 2 "Audio Workflow."

Slip mode

Figure 3.1 Slip mode selector.

This is the mode most often used for editing picture. It allows any sound to be moved to any place or track. It even allows the inattentive editor to move the audio out of sync.

Spot mode

Figure 3.2 Spot mode selector.

Spot mode is used to edit numerically to whatever time format is selected in the display. On timecode systems, this can be timecode. Editing with Spot mode allows exact placement of time stamped material. Any audio recorded in the Pro Tools system is time stamped. Imported audio can be time stamped using the Time Stamp command in the Regions List pop-up menu at the top of the Region window. This can be used to move out of sync audio back into sync.

Spot mode can also be used to place audio at an exact location when dragged from the Region List or Workspace.

Shuffle mode

Figure 3.3 Shuffle mode selector.

Shuffle mode only allows audio to be edited into continuous audio. There can be no gaps or spaces in the audio.

Shuffle mode is great for trimming sound effects with a distinct beginning, middle, and end that are too long. The middle can be edited down and the region heals itself back to continuous audio with an intact beginning and end. This mode also works well for extending backgrounds.

Grid mode

Grid mode has two settings: Absolute and Relative. Absolute has few uses in editing picture but is very useful when sounds or fades need to line up exactly with the frame lines. The grid can be set to any size and displayed with the

Figure 3.4 Grid mode selector.

"Draw grids in timeline" setting in the preferences. In Absolute Grid mode, the audio snaps to the grid. In Relative Grid mode, audio regions snap to the grid but keep their relative position to the grid. This is very handy when moving things from track to track as it will hold the region in sync unless it is moved forward or backward by the amount of the grid. With the grid set to 1 s, it's very obvious if something has been moved out of sync.

3.2 Memory locations

It is necessary to setup memory locations so that you can snap to any scene and show or hide the tracks you want to work with. The session may be rather long and contain well over 100 tracks. Memory locations are absolutely necessary to manage even a small project.

A memory location can be added from the Memory Location window or with the Enter key on the numeric keypad. The marker can be named for the scene. The yellow shield icon (which is the default) in the Memory Location window creates a visible marker in the ruler. The other icons create show-hide states, zoom states, and other functions. These can be used to create shortcuts to track show-hide states without any time marker by selecting "none" in place of the yellow shield. The entire project can now be laid out with markers to jump to any scene and any show/hide state from the Memory Location window or with the keyboard by using period (.), memory location number, period (.).

This marker system can also be used to spot ADR, Foley, and effects. However, as this will create scores of markers, it is best to create separate sessions for these edits and then import them into the master session with the "import session data" command. This will bring any selected tracks and their media into the master session. We will look more closely at this in Chapter 5 "Dialogue Editing and Replacement," and Chapter 6 "Sound Design."

The Memory Location window is used to set up markers to snap to the beginning or end of any scene and to set up show and hide states to make it possible to "zoom" in on only the tracks that are being adjusted.

Figure 3.5 The Memory Location window is used to find locations such as scene changes as well as setting up show/hide states, zoom recall, and several other settings. It is also used to spot for effects, Foley, and ADR.

3.3 The tools

Figure 3.6 The Edit tools: Trimmer, Selector, Grabber, and Smart Tool.

The three edit tools

The first tool on the left is the Trimmer tool. As the name implies, this allows the end of an audio region to be trimmed in and out. There is a second trim tool under the Trimmer tool, the Time Compression Expansion (TCE) Trimmer tool. In this case, as the end is trimmed in and out, the region is compressed or expanded to fill the new length.

The second tool is the Selector tool. This is used for selecting sections of audio.

The third tool is the Grabber tool. This is used to grab entire regions. It also becomes the finger tool when levels are displayed. The finger is used to place markers on the volume line and then move these handles to trim audio levels. Below the Grabber are the Separation Grabber tool and the Object Grabber tools. With the Separation Grabber, part of the region can be selected with the Selector tool and then moved with the Separation Grabber tool. The Object Grabber lets you select multiple regions by shift-clicking on them.

The Smart Tool

The Smart Tool links the other three tools together. It becomes the Trimmer tool when moved near the end of a region. It becomes the Grabber tool when moved below the waveform plot. The cursor becomes the Selector tool when moved above the waveform plot. It also becomes an auto fade tool when moved to the top or bottom of the end of the region.

In Pro Tools, there are few editing tools. When the editors who have always had a huge kit of razor blade tools, extending tools, and selecting tools see basically three tools in Pro Tools, they are lost. But these three well-chosen tools, which can be linked into one Smart Tool, can make any edit. This is great for the editor – there's no need to constantly go for a new tool wasting time in tiny little increments that really add up by the end of the day. Although it does take a little time to master "one tool" editing, it's fast, accurate, and very capable.

Keyboard Tools

Keyboard Focus – Link Timeline and Edit Selection selectors

Figure 3.7 In Pro Tools 8 the Zoom toggle, MIDI Mirroring, Tab to Transients, Link Timeline and Edit Selection (selected) Link Region and Timeline and (moved from preferences) the Edit Insertion Follows Playback buttons are bigger and easy to find. The keyboard focus for the edit window is dropped, but still used in the edit groups and Region List windows.

The "focus" of the keyboard can be moved from window to window. The "a...z" indicator is selected when a blue box appears around it. There are three such indicators in the Region List, Mix groups list, and Edit groups list. When selected, the keyboard can be used to search that window by typing the first letter of the desired region or group.

There are several tools in the Edit menu that can be used much faster from the keyboard. Several are boilerplate functions used by most software.

Function	Macintosh/Windows
Cut	Command/Control x
Copy	Command/Control c
Paste	Command/Control v
Clear	Command/Control b
Select All	Command/Control a
Separate (razor)	Command/Control e
Undo	Command/Control z

The Link Timeline and Edit Selection selector links any audio region or selected area as the playback in and out points.

When selected, the Tab to Transients tool will jump to the next sudden sound when the tab key is pressed.

The Nudge command

In Pro Tools 8, the Nudge indicator is displayed at the top of the edit window in a new large window that it shares with the Grid setting. Both now read in minutes and seconds as well as timecode, even if you do not have the time-code option. The Nudge command will nudge any selected audio region, or if none are selected, it will nudge the insertion point by the distance selected in the Nudge indicator. The Nudge value is controlled by the + and − keys on the numeric keypad. Like many other edit tools, Nudge even works during playback, great for syncing errant sync.

3.4 Editing to video

When editing to video, the goal is to align the audio regions to the picture. When an audio region is moved around, the display shows the frame corresponding to the beginning of the audio region. Unless the region is being placed to timecode, it will be slipped into sync visually. Therefore, the simplest system involves trimming away any of the region before any wanted sound. Then, slip the region back and forth until the beginning of the audio is in the proper location as seen on the video screen. Now find the end of the region. If the region is too long, select all of the middle of the region and press the Delete key. Slide the end of the sound into the proper location. The Trimmer tool shows the video frame that corresponds to the end of the region. Slide the region into place with the Grabber and trim the end to picture with the Trimmer tool. Now fill the middle by extending either the head region or the tail region with the Trimmer tool. The missing middle can be duplicated into several subregions if the sound is too short. If necessary, you can crossfade the two regions with an automated fade. This can be done with the Smart

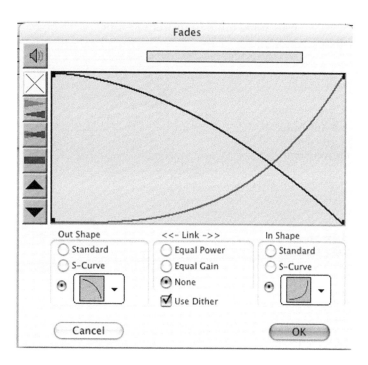

Figure 3.8 The Fade editor is used to alter the speed and ramp or fade ins and outs.

Tool at the edit point or by selecting the area for the crossfade with the Selector tool and pressing f. The fade can be edited by double clicking on it.

Another slick trick is using the TCE Trimmer tool under the standard trim tool. Now the audio is moved into position with the Grabber; the video display showing the beginning of the audio region. Now move to the end of the region using the TCE Trimmer tool, and trim to the point on the video where the sound should end. The video display will show the end frame of the region. The region will now compress or expand to fit. This system only works if the length of the region was already somewhat close. Otherwise there are horrendous artifacts.

The Pencil tool

The Pencil tool can be used to draw automation into volume overlays or to draw new waveform in place of the original. This involves zooming in to the sample level. Then a new waveform can be drawn to replace an unwanted sound on the original track. As this is destructive, the region should be duplicated with the AudioSuite Duplicate plug-in before drawing over the waveform. It's very easy to draw in a sound much worse than the one you are attempting to remove. One wonders if a million chimpanzees drew waveforms for a million years if one of them would draw Beethoven's Ninth, or at least some tome from Trent Reznor.

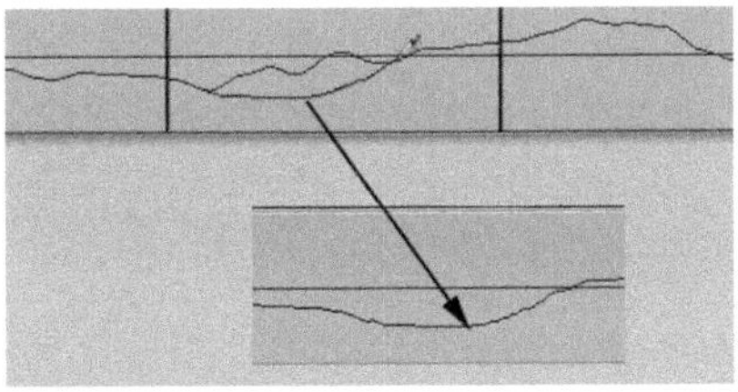

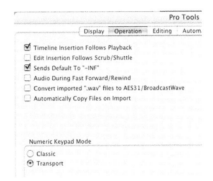

Figure 3.9 The Pencil tool can be used to redraw the waveform.

Scrolling

Figure 3.10 The Preferences – Operation window. In this case with the "Timeline insertion follows playback," when playback stops the insertion point moves to the stop point, at times a very handy mode – so handy that this is now on the edit window as a button in version 8.

There are various scrolling options, almost as many as there are editing tools. On all systems, there are No Scrolling, Scroll After Playback, and Page scrolling. In all cases, the insertion point stays fixed and the playback snaps back to the insertion point when playback resumes. This can be very handy but it can also be problematic.

On HD systems, there are also the Continuous Scrolling with Playhead and Continuous Scrolling settings. In these modes, the head stays fixed in the center of the screen and the media scrolls underneath it. In the Continuous Scrolling with Playhead mode, when you stop playback, the playhead and media stay in the current location even though the insertion point remains in the same place. This can be a very handy function, but there is no equivalent function on LE and M-Powered systems. There is, however, a handy selector in the edit window (found in the preference settings in version 7 in the Preference-Operation settings) called "Timeline insertion follows playback."

3.5 The mix window

Figure 3.11 Mixers can be large imposing pieces of equipment. But their function is actually rather simple.

The fundamental tool of the dub is the mixing board. As the name implies, this device combines sounds together. The board can be quite imposing with hundreds of knobs, sliders, buttons, and hundreds of flashing lights, yet the fundamental function is rather simple.

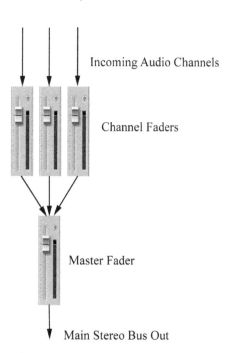

Figure 3.12 A basic three channel to stereo mixer.

The most basic version has two or three inputs and a single mono output. Three knobs control the level of the sounds from the three inputs and a forth knob controls the level of the output. This is a simple way to mix several sounds together while altering their level over time. A more complex version may be fundamentally the same, but with more inputs. If the mixer is expanded to 24 channels, now 24 sounds can be mixed in real time to a single mono sound. Several problems arise when we expand the mixer to 24 channels. First, 24 knobs are unmanageable unless you have 24 hands. Sliders work much better on a larger mixer as several sliders can be manipulated with each hand.

Automation also helps when attempting to control a large number of sliders. Each slider can be programmed to play at a certain level at a certain time, freeing the operator (mixer) to manipulate other channels while the programmed sliders adjust themselves.

Adding more output channels can further expand the mixing board. If the board is expanded to have two output (stereo) channels, the input channels must now have a way to set the output to the desired output channel. Moreover, the input channels themselves can now be stereo channels, effectively two input channels controlled by a single slider. The most common way to assign the channel to one of the output channels is with a pan pot, a second slider that moves the sound right, left, or anywhere in the middle. This slider is usually very small and set horizontally to indicate it is a pan and not a level set. It may also be a knob. The output could also be set with a button or channel assign. By pressing the right or left button, the channel can be assigned to right, left, or both. Although the pan pot makes more sense for the stereo mixing board, the channel assign button makes more sense on very large mixers with many output channels.

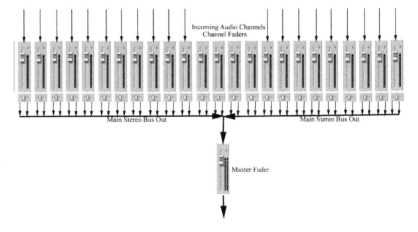

Figure 3.13 Even though this hypothetical mixer is so far rather large, it's still very simple. There are 24 automated mono sliders and 24 pan pots feeding a stereo automated output.

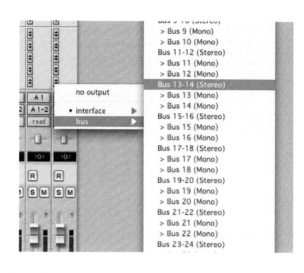

Figure 3.14 The outputs need not be sent to the main stereo bus. If the mixer has aux busses, the channels can be sent to an aux bus rather than the main stereo bus. This will require bringing the aux bus back to the main stereo bus for output. Even the smallest Pro Tools systems have 32 mono (16 stereo) busses.

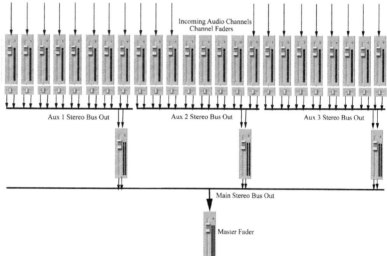

Figure 3.15 The output of any channel can be assigned to an aux bus rather than the main stereo bus. A stereo Aux channel can then be set to this bus as its input and the main stereo as its output. These are great for stem mixes, one each for dialogue, effects, Foley, and music.

A stereo Aux channel needs to be created to serve as the return for the aux bus to the stereo main bus. On the Pro Tools mixer, this is done by calling up New Track in the Track menu and selecting stereo aux input.

By using the aux buses, it is possible to add complete separate mixers to the board. If we add a second mixer to the audio path before the main slider, we now have two mixers each feeding a different output. These are called auxiliary mixers. If the main sliders do not affect them, they are referred to as prefader mixers. If the level of the main fader controls the volume of the input to the aux mixer, the mixer is said to be a postfader aux mixer. Because sliders are large and take up a lot of space, aux mixer level controllers on hardware mixers are usually knobs. Each aux mixer will also have a master output controller. It is common to have several aux mixers, on large systems perhaps as many as 12.

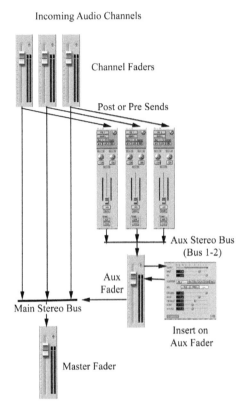

Incoming Audio Channels

Channel Faders

Post or Pre Sends

Aux Stereo Bus
(Bus 1-2)

Aux
Fader

Main Stereo Bus

Insert on
Aux Fader

Master Fader

Figure 3.16 Effects send/return setup.

Aux mixers can be used to send a separate mix to some piece of equipment, recorder, monitor speakers, or whatever the mixing strategy calls for. If the aux mixer is sending its mix to a piece of equipment, this audio will need to be returned to the main mix. So the board will also have several returns, channels intended simply as a way to return an aux mix back to the main mix.

Every channel will also have inserts. The simplest version of this is a jack for every channel on the back or even the front of the board. An insert cable can be plugged into this jack. The insert cable has a Y configuration, two cables coming from one plug. One cable is the input cable and the other the output. With this, any device can be cabled into the channel and the audio will be routed through the outboard device and returned.

On the Pro Tools mixer, the inserts are above the Aux sends in the darker gray area. Normally, plug-ins are inserted. Hardware can also be inserted by setting up one or more of the channels on the audio interface as inserts in the IO setup window. There will be more on this in Chapter 8 "The Dub."

As equalization is almost always needed on every channel, rather than tie up an insert adding an outboard equalizer (EQ) to every channel, most hardware boards have some basic EQ built into every channel. This may be three or four band tuneable equalization. In Pro Tools, this is an insert attached when needed.

Each channel will also have a mute button and a solo button. The mute mutes the channel so that it does not sound. The solo mutes all tracks other than the soloed track. Any number of tracks can be muted or soloed.

Our "basic" mixer is now rather imposing with 24 channel sliders, 144 aux mixer knobs feeding six 24 channel aux mixers, 144 EQ knobs controlling 24 four band EQs, 24 pan pots, 24 solo buttons, 24 mute buttons, an output slider, perhaps six returns, six aux output master knobs and dozens of buttons to control automation, assigns, and pre-post controls for the aux mixers. Yet in this forest of controls, it is still very straightforward.

Figure 3.17 The virtual Pro Tools mixer.

The virtual Pro Tools mixer can also be imposing, yet it is too basically simple and easy to understand. And because it is virtual, it is customizable. Inserts and submixers are only added when and if they are needed, and the type of EQ is not fixed as part of the design, making it possible to use any number of different EQs, mix and match them, add compressors, gates, expanders, reverbs, delays, and any number of other devices as inserts with no need to patch or use any physical devices. Aux mixers can be created as the need arises, as can returns.

Although it may be tempting to simply add any needed plug-ins, any mix has a way of becoming rather large and management of resources is always important. Moreover, controlling scores of plug-ins is time consuming and difficult. A better system is to route groups of tracks through a single plug-in by routing through an aux fader. This kind of "send/return loop" is simple to set up, simple to use, and uses fewer system resources than adding plug-ins to every channel.

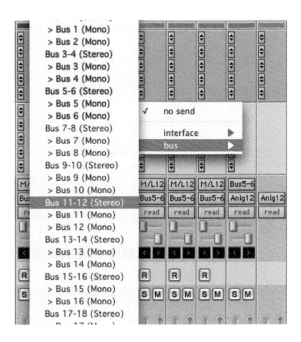

Figure 3.18 Up to 10 sends can be sent to the busses from each channel.

Multiple sends can be added to any track in the sends area of that channel. There are ten for each channel, displayed in two groups of five. The outputs of these sends are then routed to an unused pair of bus channels. Depending on the size of the Pro Tools system, there are at least 32 such busses. The aux fader on the output of the three channel faders is created in much the same manner. In this case, an Aux channel is created and the input set to any unused pair of bus channels. The outputs of the channel faders are also set to this pair of bus channels.

Each channel that is to receive the effect of the plug-in is given an aux send fader. As in the illustration, usually this is a postfader send so that the channel slider still controls the total volume level of the channel. The outputs of the sends are routed to an Aux channel and from there to the main stereo bus. Any desired effects can be inserted into the return aux fader. In the illustration, another aux fader has been added as the output of the three original faders so that it becomes easy to control the balance between the effects return and the original sound.

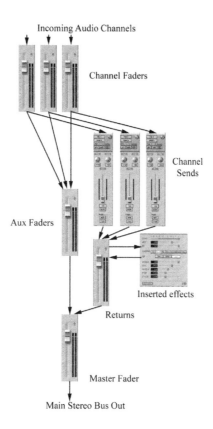

Incoming Audio Channels

Channel Faders

Channel Sends

Aux Faders

Inserted effects

Returns

Master Fader

Main Stereo Bus Out

Figure 3.19 The send/return loop with aux submix fader from channel.

Figure 3.20 In this example, an RTAS AIR Kill has been added to the dialogue (DX) channels. There is also a DX subfader for the music tracks both routed to a stereo master fader.

A common use for such a setup would be to add a reverb to use in any scenes where reverb is called for. The reverb can be added when needed by automating the mute button or even the volume slider on the return aux fader. The amount of reverb for each channel is controlled by the individual channel sends, which can also be automated.

In film postproduction, it is also common to mix to final stems; the mixing board is often set up to simultaneously lay the music, effects, and dialogue stems off to another channel or channels. In most cases, they will be recorded or exported to delivery stems after the final mix. With 5.1 and other multitrack finishing formats, there may be 5–6 music stems, 5–6 effects stems, and 3–5 dialogue stems. Stems for a stereo mix are usually one or two dialogue, two (stereo) music, and two (stereo) sound effects. The stems can be used for editing the trailer, "dubbing" into a foreign language, or even remixing for video. Stems are an absolutely necessary asset of the motion picture.

To set up the board for stem mixing, simply route all the effects tracks to a FX aux submaster, all music to a MX aux submaster, and all the dialogue to a DX aux submaster. This also allows for tight fine-tuning of the final mix by tweaking the stem faders to optimize the balance among the music, effects, and dialogue.

On very large projects, it is often necessary to connect several Pro Tools systems together. Although one system can be incredibly powerful, in postproduction is often necessary to have several people at the mixing board at the same time. Three and four person mixes are not uncommon.

Up to five Pro Tools HD systems can be interlocked with the Satellite Link option, or any number of systems can be interlocked together via timecode and SYNC HD. This allows each mixer to have control over their system, with the exception of transport control, which must be controlled by one mixer. Usually this is the dialogue mixer. Each system can have up to four computer screens; however, each system can only have one mix window open. By linking several systems, each mixer has their own mix window displaying the tracks on their system. The master system has inputs from the other systems with aux faders controlling these inputs and routing to the proper stems that are also on the master system.

It is also common to use another Pro Tools system as an insert recorder in the final mix. In the "old days," over 15 years ago, sound was usually dubbed on a dub stage from "dubbers." These playback machines were interlocked with a projector and an insert recorder. Before 1950, these recorders were optical and not able to insert, back up, or record over a previous recording. With advent of magnetic recording, there soon followed the insert recorder. This recorder could "punch in" over an existing recording without making any tick or pop. If a section of the mix was not right, the entire interlock could be backed up, rolled forward, and the recorder could be placed into record, the section remixed and the recorder taken back out of record.

It's hard to believe, but early films such as *The Wizard of Oz* and *Gone With the Wind* were mixed in real time in one pass. Fortunately, they were mixed in 10-min reels, but if there was any mistake in the 10-min section, it meant starting all over.

These days it is much more common to spend weeks adjusting the automation of the mix and then simply "bounce to disc," in other words record the automated mix in real time to a drive.

But some mixers feel something is lost by not performing at least some part of the mix live, and moreover they feel they are faster when mixing in real time than when setting up elaborate automation. In this case, a separate Pro Tools system interlocked to the master system can function as the "insert recorder." This requires a monitor controller system known as a "pec/direct." The mixer(s) need a way to listen to the playback from the recorded mix, compare this to the output from the mixing board, and when all the levels match, punch into record. Most studios purchase or construct a button panel to switch from the output of the insert Pro Tools system to the output of the playback Pro Tools system(s). The insert system is set up to record stems and the composite mix at the same time. This way the drive of the insert system ends up containing the deliverable finished tracks. Pec refers to "photo electric cell," a last vestige of the old optical dubbers. The pec/direct selector switches the monitors from the output of the mixing console and the playback of the recorder, referred to here as the pec.

The huge advantage of the virtual mixer is its flexibility and able to be customized to the needs of the project as well as the mixing style of the mixers. And with hundreds of powerful plug-ins available, the powers of racks full of hardware are at the mixers' fingertips without the need for the racks or the extra air conditioning.

In this chapter

Plug-ins

4.1 Types of plug-ins

Plug-ins come in several types. RTAS (Real Time AudioSuite) plug-ins will work in any Pro Tools system, HD M-Powered, or LE. They use the computer's processor and memory, requiring no special hardware. There are also TDM plug-ins. These will only work on HD systems as they use the DSP chips in the HD boards to process audio. Both types function as inserts that are inserted into an audio path and process audio in real time as the audio moves through the plug-in.

There are also AudioSuite plug-ins for processing audio to new media files. These are used from the AudioSuite menu in Pro Tools.

Within these types, there are many other categories of plug-ins, filters, processors, gates-compressors, virtual musical instruments, and audio routing software. There are hundreds of plug-ins available from Digidesign as well as other suppliers. As it is possible to spend hundreds of thousands of dollars on plug-ins, the question comes up, what do I really need?

Most of the plug-ins available are intended for music production and even live concert mixing. Although it's impossible for one person to pick plug-ins for someone else's creative use, there are several plug-ins that anyone editing and mixing for cinema should consider.

Many of these plug-ins may be something you will only use once or twice in a lifetime, but the good news is that many can be rented for a day or two on the Digidesign Web site.

Most plug-ins are iLok protected. The plug-in license is loaded into the iLok plastic key that must be plugged into a USB port before the plug-in can be used. The Pace iLok is available in many computer stores and online.

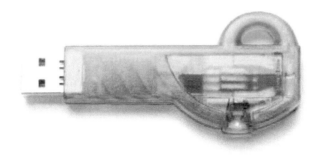

Figure 4.1 The Pace iLok available in stores and online.

Let's take a closer look at some very useful plug-ins. Links to dealers can be found on the Digidesign Web site.

4.2 Video tools

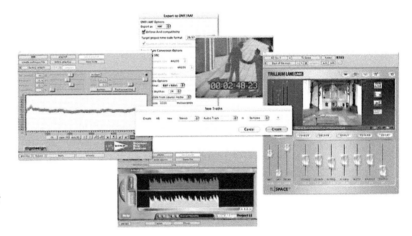

Figure 4.2 The DV Toolkit is a bundle of selected plug-ins and tools that are excellent for working with video and film.

DV Toolkit 2

Perhaps the most important basic plug-in set used in film and video is the DV Toolkit. This is not a plug-in but several options and plug-ins sold as a set. Although very expensive, these options and plug-ins would cost even more if purchased separately. DV Toolkit 2 includes these options and plug-ins:

- DigiTranslator 2.0 for import/export of OMF, AAF, and MXF files
- TL Space Native Edition convolution reverb

- Synchro Arts VocALign Project time-alignment tool

- Digidesign Intelligent Noise Reduction (DINR) LE noise reduction plug-in

The DV Toolkit also allows LE users (not M-Powered) to use some features only available on HD systems:

- Timebase rulers (Timecode and Feet + Frames)

- Timecode Rate selector

- Feet + Frame Rate selector

- Current Timecode Position command

- Current Feet + Frames Position command

- Use Subframes option

- Audio Rate Pull-Up and Pull-Down

- Video Rate Pull-Up and Pull-Down

- Timecode Mapping options when importing tracks

- Custom Shuttle Lock Speed

- Scrubber Tool for scrubbing the Video track

- DigiBase Pro, the full-featured version of the DigiBase file management tool

- Timecode and Feet + Frames functions

- Powerful editing and session management features

- Up to 48 mono or stereo tracks at up to 96 kHz on hardware interfaces that can support these sample rates

- Pro Tools MP3 Option

The most important plug-in in this set is the DigiTranslator. We looked at this plug-in in Chapter 2 when we were discussing audio workflow and importing OMF files. Although converting OMF files to Pro Tools sessions is the most common use for DigiTranslator, it is capable of also importing AAF and MXF as well. This allows the sound edits from picture editing systems like Avid and Final Cut Pro to be imported into Pro Tools. You don't use this function very often; however, when you need it you need it; the workarounds are so cumbersome, and DigiTranslator is simply necessary. Although the rest of the DV Toolkit is intended for LE systems, DigiTranslator is a must for HD systems as well.

TL Space Native Edition convolution reverb includes a comprehensive library of high-quality sampled rooms and spaces as well as reverb effects. This can be used to make wireless mics and ADR sound like "live" recordings in a certain space. These spaces are also tunable. Although this is AudioSuite only, HD users may want to use it as well.

VocALign Project is used to pull ADR and wild lines into sync with a guide track. If the rerecorded line is at all close, VocALign will pull it into sync. The DV Toolkit includes the LE version, which is very powerful, but there is also an HD version.

The DINR LE noise reduction plug-in is old as dirt and just as solid. It is used by sampling a small section of noise into DINR, which becomes a tuned noise gate that treats the sampled noise as nonsignal. The result is good removal of the sampled noise. There is an HD version as well and a set of TDM HD only noise reduction tools also known as DINR that includes this plug-in.

The timecode, pull-up and pull-down, and DigiBase options bring these post-production tools, available on all HD systems, into LE versions.

Music Production Toolkit

Why, you may be asking, is the Music Production Toolkit here with video tools? When the Music Production Toolkit and DV Toolkit are bought together, they are known as the Complete Production Toolkit for Pro Tools 8 LE only. The Music Production Toolkit adds the Digidesign Hybrid synthesizer. This is one of the best early sounding synthesizers combined with a great sample player. It needs a Musical Instrument Digital Interface (MIDI) keyboard to be very useful. The Music Production Toolkit also adds SoundReplacer and Smack! compressor.

When bought as the Complete Production Toolkit, Pro Tools 8 LE users also get 5.1 mixing with the full 5.1 surround controls from the HD systems. This also requires one of the 003 audio IO devices, but the end result is an LE system that is more of a mini HD system.

4.3 Reverbs and convolution reverbs

Reverbs and convolution reverbs are used to add room acoustics to "dry" audio, be it effects or dialogue. A reverb adds in a decay causing the audio to "trail off." This should not be confused with a delay that actually holds the audio off and then plays it later. This can be used to add an echo effect and can be used in conjunction with a reverb. Reverb is much faster than delay echo and has no discernable echo.

Convolution reverbs use digital samples of actual room acoustics recorded, analyzed using a mathematical convolution operation, and then converted into a digital reverb preset. It is even possible to record the output of old spring and oil-filled reverbs from the 1960s and convert these into convolution presets. The best of these are smooth and natural with no electronic feel. Sample sets are also available for these plug-ins, expanding the sonic possibilities.

One of the biggest problems in selecting the right reverb is the sheer number available today. There are scores on the market and all are good, and many are great.

Digital Reverbs

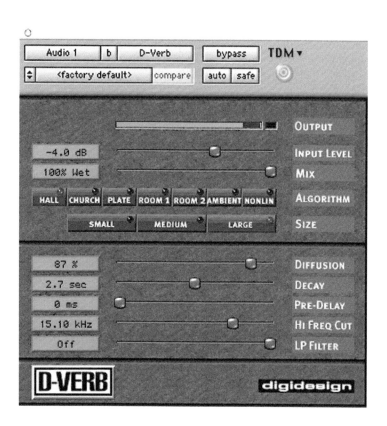

Figure 4.3 Digi's free D-Verb plug-in is a good basic digital reverb.

One of the top features of the Digidesign D-Verb is the price: free. Yet this little digital reverb is quite capable and adequate in many situations. Download it from the Digidesign Web site.

TrueVerb

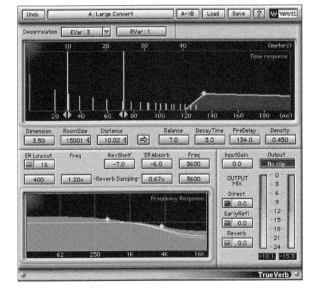

Figure 4.4 Waves TrueVerb is a classic digital reverb with graphic modeling of the room parameters. It also features preverb or early reflections that are common in actual room acoustics.

Reverb One

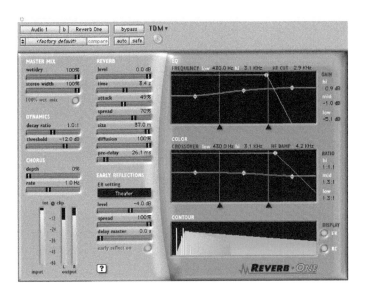

Figure 4.5 Reverb One from Digidesign.

Reverb One from Digidesign is a first-class TDM reverb that can hold its own in any situation. The only drawback is that there is no native RTAS version.

Features:

- Independently controllable reverb settings, including level, decay time, attack, spread, room size, diffusion, and predelay

- Extensive library of reverb presets

- Early reflection controls, including level, delay, and spread, plus early reflection room presets

- Dynamics and chorus sections for shaping reverb decay and optimizing reverb effects

Convolution Reverbs

TL Space

Figure 4.6 TL Space convolution reverb for music and postproduction.

TL Space is a good, solid Pro Tools convolution reverb for music and postproduction applications. If you are only going to have one reverb in your kit, consider this one. By combining the sampled convolution acoustics of real spaces with DSP algorithms, TL Space offers stunning realism with full tweakability of the reverb parameters in both mono and stereo formats. A TDM version is also available.

TL Space includes a comprehensive library of high-quality sampled reverb spaces and effects ideal for music and postapplications, and you can download additional impulse responses from the TL Space Impulse Response Library.

Features:

- Supports RTAS and AudioSuite processing

- Fully featured reverb engine for mono and stereo formats

- Includes a library of reverb and effect impulses for music and post applications

- Impulse library browser for quick and easy auditioning

Altiverb

Figure 4.7 Often voted "best reverb ever," Altiverb is the original convolution reverb.

Introduced in 2001, Altiverb is the original convolution reverb. Altiverb is the favorite of many mixers because of its features and its ever-growing library of acoustics samples. It is often voted "best reverb ever" by magazine reviewers as well as top sound editors and mixers.

Features:

- TDM and Native processing

- Stage positioner – put your instrument anywhere on stage

- 40 snapshot memories for automatable total recall

Acoustics samples include the Sydney Opera House, Notre Dame Cathedral, Cello Studios LA, 20 car interiors, caves, forests, lots of essential classic reverb gear, a Jumbo Jet cockpit, and much, much more.

Waves IR-1 Parametric Convolution Reverb

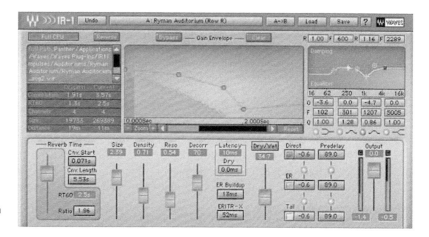

Figure 4.8 Waves IR-1 Parametric Convolution stereo reverb.

The Waves IR-1 Parametric Convolution Reverb features many sampled rooms including the Wembley Arena, Grand Ole Opry, Sydney Opera House, Birdland, and over a hundred more. Fully tuneable and tweakable, the IR-1 is an amazing stereo reverb. Waves also offer the IR-L Convolution Reverb.

Waves IR-360 Parametric Convolution Reverb

The IR-360 takes the Parametric Convolution Reverb into the surround theater. The room sampling is now 360° and still fully tuneable.

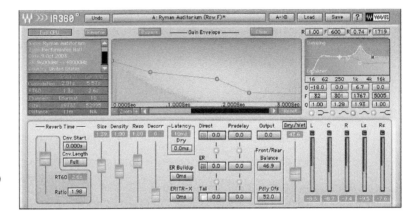

Figure 4.9 Waves IR-360 Parametric Convolution Reverb.

4.4 Futzing plug-ins

Often the goal in postproduction is not to make the sound quality great; it is to make it right. The sound may need to sound like it is coming through a telephone, an old jukebox or radio, or over a bullhorn. There are several great plug-ins for mutating, twisting, and generally ruining the audio until it is perfect.

SpeakerPhone

Although this is a "phone futz," it has presets for speakers of all sizes as well as their environments. Although these backgrounds are good, they should not be added to the mix by a plug-in. Backgrounds should be placed in background tracks. However, this plug-in has digital recordings of everything from feedback to dial tone that can be recorded into a background track.

Speakerphone uses actual samples of hundreds of original speakers, a radio tuning dial, record player scratch and static, GSM phone compression, distortion, tremolo, delay, EQ and dynamics, bit crushing, sample rate reduction, and a full-blown convolution reverb.

D-Fi

D-Fi consists of four unique plug-ins: Lo-Fi, Sci-Fi, Recti-Fi, and Vari-Fi. Available for TDM, RTAS, and AudioSuite, these plug-ins create retro synth distortions reminiscent of the 1970s and 1980s. Lo-Fi provides bit-reduction for retro sound processing. Sci-Fi adds analog synth-type ring modulation, frequency modulation, and variable frequency resonators – Indispensable if you

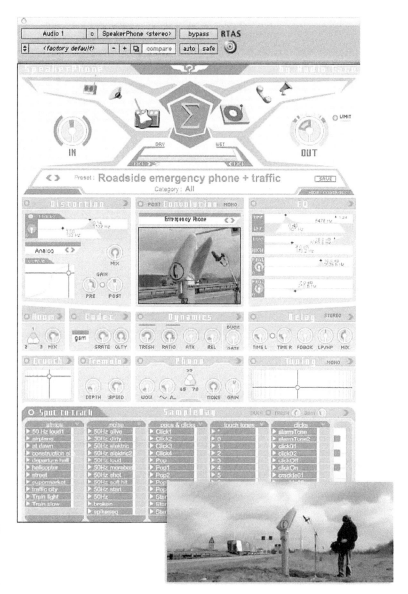

Figure 4.10 SpeakerPhone is used to reproduce the sound quality of speakers and telephones. It even produces background noise.

are remaking *Battlestar Galactica*. Recti-Fi provides super- and sub-harmonic synthesis, and Vari-Fi creates the effect of analog audio changing speed to or from a complete stop. Although you may go a lifetime without ever needing these esoteric distortions, if the need arises the set can be rented for 2 days on the Digidesign Web site for $10.

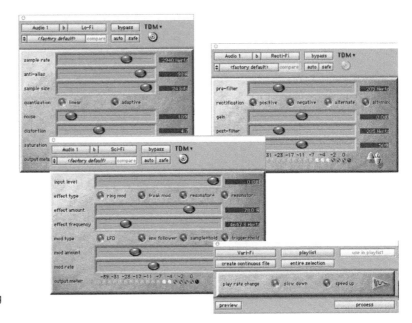

Figure 4.11 D-Fi is used to futz audio into weird and noisy analog-seeming sound.

Cosmonaut Voice

Figure 4.12 Cosmonaut Voice from the Bomb Factory.

Top secret documents recently declassified from the files of the KGB (Komitijet Gosudarstvjennoj Bjezopasnosti) have brought us Cosmonaut Voice. Quoting from the original user manual: "Superior advanced space technology by great scientists of Union of Soviet Socialist Republics brings creation of Cosmonaut Voice advanced communication simulation. Production of voices from the cosmos and beyond is easy as cake. Simply turn knob intended for addition of noise (πøpør) when level of comrade actors voice reaches certain level as defined by rotation of threshold knob (IIIYM). This triggers advanced simulation of Soviet squelch sound or inferior American beep tone used by NASA who claims to have placed flag on moon." Certainly a plug-in that sounds as bad as it looks. But if crappy 1960s radio or walkie-talkie sounds are called for, here you go! It even does a wonderfully bad cell phone. Can you hear me now?

Trash by isotope

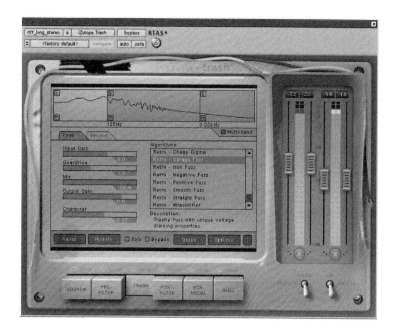

Figure 4.13 Trash by isotope. When all else fails, take out the trash.

The name about sums it up: Trash. You can make it buzz, fuzz, squash, and trash. A very high-tech way to sound low-tech. This plug-in is intended for use on heavy metal music to make it feel more over-driven and distorted, but don't rule out a crappy PA system. And don't worry about the loose wires, the guy is coming to fix it.

The AIR tools, free with Pro Tools 8, are great for messing up dialogue to make it "just right." Try the Fuzz Wah as a phone futz. Awesome! Or the Talkbox for robots or even Eleven as a PA system. The AIR D-Fi is much better than the old TDM version, and it's free. You can't beat that. Some of these tools are good for sound design as well, but they shine as dialogue futzes. Oh, and I guess they work for heavy metal music as well.

Figure 4.14 New to Pro Tools 8 are the AIR (Advanced Instruments Research group) tools, often used with the AIR instruments, these can also be used to great effect on vocal tracks to "futz" the sound.

4.5 Surround sound tools

Surround sound comes in many formats from several companies. Most of the theatrical formats require encoding by technicians from the companies that provide and license the use of the formats. Most of these formats also have a DVD version that can be very similar to the theatrical version. These too have required licensing and encoding by the companies like Dolby. However, with the explosion of DVDs and small-run DVD reproduction, some formats have become available for self-encoding without the need for a specific one time license, hardware, or a visit from a company technician for the project.

5.1 also often requires bass management for many Pro Tools systems if the Low Frequency Effects (LFE) channel is also going to be used as a satellite sub-woofer. (This is covered at length in Chapter 1 "Pro Tools Systems.")

The most popular surround formats are 5.1 and LCRS (Left Center Right Surround). 5.1 refers to the number of channels, in this case five full range channels, left, center, right, left surround, and right surround. The point one refers to an LFE channel. In LCRS, there are only these four channels; however, very low frequencies can still be routed to the LFE through a crossover system.

Dolby Surround Tools

Dolby Surround Tools encodes the four-channel Dolby LCRS format. This is an analog format that uses phase angles to encode stereo audio into the four LCRS channels. There are several advantages to this format. First, it is analog stereo audio and is fully compatible with analog stereo. If it is not decoded into LCRS, it plays normally in stereo. It is the most basic surround format and has been used for years.

The disadvantages are that it is falling into disuse as all digital formats have become the norm. It also requires adding the phasing matrix as the project is mixed, while digital formats can be encoded after the mix. Dolby Surround Tools adds in this matrix without the need for any hardware.

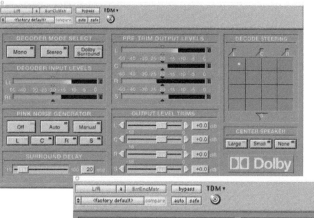

Figure 4.15 Dolby Surround Tools is used with the Dolby LCRS format.

Dolby Surround Tools also comes with a pink noise generator and decoding system for calibrating the system and decoding in playback.

SurCode for Dolby Pro Logic II and Neyrinck SoundCode for Dolby Digital and DTS

These three plug-ins from SurCode and Neyrinck encode Dolby Digital, Dolby Pro Logic, and DTS. They also allow for playback of these encodes as well as finished DVDs for quality assurance.

SurCode is used to encode Dolby Pro Logic II for games and DVD. Pro Logic II is similar to Dolby LCRS but is a 5.1 (six-channel) format. Like LCRS the multiple channels are encoded into stereo audio that is 100% stereo compatible. Comprehensive monitoring allows for auditioning of both the source and encoded audio. Virtually every Dolby Digital home theatre system sold today can decode this Pro Logic II stereo. It is also backwards-compatible with Dolby Pro Logic receivers and decoders.

Neyrinck SoundCode for Dolby Digital plug-in suite provides tools that enable you to encode and decode Dolby Digital (AC-3) audio directly within Pro Tools

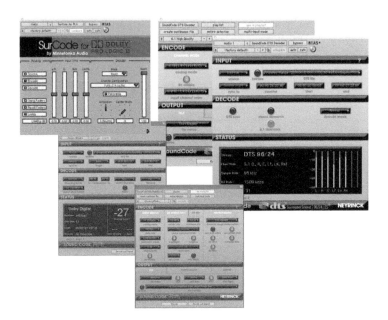

Figure 4.16 SurCode for Dolby Pro Logic II and Neyrinck SoundCode for Dolby Digital and DTS.

software, without the need for dedicated Dolby hardware. Although AC-3 files can be created with applications like Apple's compressor, SoundCode for Dolby Digital performs this encode right inside Pro Tools and provides for playback of the encoded audio. Dolby encodes compress the sound and can be heard flattening the audio. Because there is loss in the Dolby encode, the quality of the encode is critical. These plug-ins produce high-quality encodes, where compression software may not do the job as well.

Similar to SoundCode for Dolby Digital, Neyrinck SoundCode for DTS plug-in enables surround encoding and decoding of the DTS DVD format directly within Pro Tools.

Waves 360° Surround Bundle

This is expensive but an absolutely necessary set of surround tools. 360° Surround Tools includes compression, limiting, reverb, spatial enhancement, bass management, and much more.

It includes:

- S360° Surround Panner
- S360° Surround Imager
- R360° Surround Reverb
- C360° Compressor

Figure 4.17 Waves 360°
Surround Tools Bundle.

- L360° Surround Limiter
- M360° Surround Manager
- M360° Surround Mixdown
- LFE360° Low Pass Filter
- IDR360° Surround Bit Re-Quantizer.

ALR (Audio Research Labs) Sound Stage

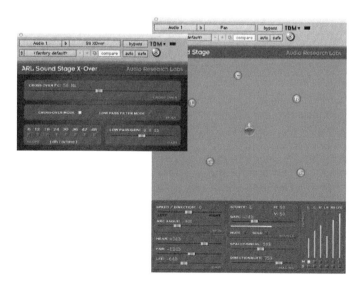

Figure 4.18 ALR Sound
Stage.

ALR Sound Stage provides some of the features found in the Waves 360 set at a much lower price. The ALR Sound Stage X-Over provides the much needed bass management on Pro Tools systems that do not have full range speakers. ALR Pan provides spatial enhancement, excellent stereo-to-5.1 conversion, and some interesting panning functions for moving the entire surround prospective around the listener.

SurroundScope

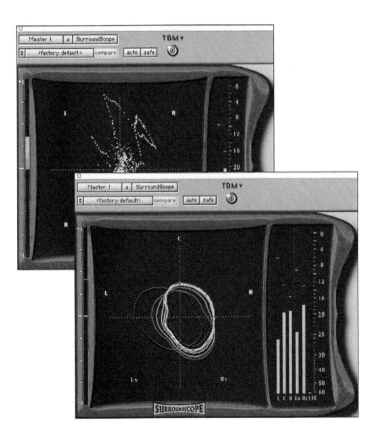

Figure 4.19
SurroundScope, a free
DigiRack Plug-in.

SurroundScope provides a graphical display of the signal level for each audio channel within the multichannel sound field. You can visually monitor what your listeners will hear. As the perceived location is highly dependent on phase angles, SurroundScope comes with a full-featured Lissajous Phase Meter, a simple yet highly accurate way of monitoring the phase of your stereo mix or other stereo tracks in your session.

Features:

- Real-time display of stereo or multichannel track data

- Support for all standard multichannel formats

- Automatic detection of multichannel track formats

- Multidimensional display depicts the position of the audio signal within your multichannel track/surround mix

SurroundScope is now part of the free SignalTools DigiRack Plug-in for Pro Tools 7.0 and higher HD systems.

4.6 Sound design

Often the needs of the sound designer require being able to turn donkeys into dragons and cats into demons. There are many plug-ins able to twist audio into something truly unworldly or over worldly if needs be.

Pitch shifters and time expansion and compression

The most basic tools in the sound designer's kit are pitch shifters and time expansion and compression. Slow down a shopping cart and it becomes a movable gorilla cage. Pitch it down and it becomes a rumbling war machine. There are many great tools for pitch shifting and time compression, some good basic tools even come with your Pro Tools software for free.

Although definitely not free, Pitch 'n Time is one of the most powerful pitching and time compression tools available.

Since Pitch 'n Time my life is fine!

David Lynch, Film Director

Features:

- Modify speed from 1/8th to eight times the original rate and simultaneously pitch-shift by three octaves

- Preview changes in real time

- Unrivaled processing quality

- No loss of timing accuracy

- Process multichannel audio

- Process Dolby matrix-encoded tracks without losing surround information

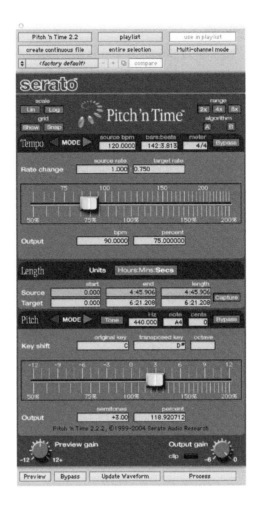

Figure 4.20 Pitch 'n Time by Serato.

SoundToys TDM

You are never too old to play with toys. SoundToys is great and simple to "play with" when the goal is creating effects for mixing, sound design, remix, post, and audio for gaming. It has nine plug-ins and thousands of effects. There is a limited RTAS version as well.

- EchoBoy – Pro analog echo/delay modeler, 30 vintage and modern models and MIDI sync

- SoundBlender – Industry standard, powerful multieffects plus intelligent pitch-shifting

- FilterFreak – Fat analog sound, flexible modulation, MIDI sync'd rhythm

- PitchDoctor – Smooth, easy, auto vocal pitch correction without "chipmunks"

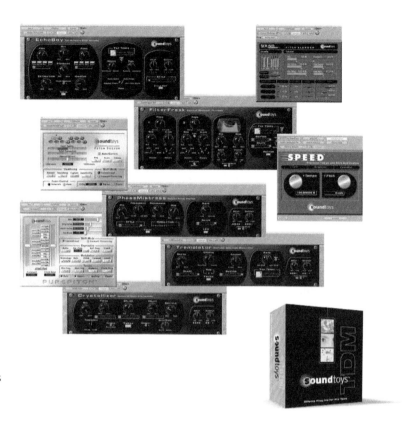

Figure 4.21 SoundToys
Who say's work can't be
fun?

- PhaseMistress – Lush analog-sounding phase-shift, six mod modes, MIDI sync'd rhythms
- SPEED – Pitch and time compression/expansion tool
- Tremolator – Classic tremolo plus programmable auto-gate with MIDI sync
- PurePitch – Formant and pitch shifter, harmony, and voice manipulation
- Crystallizer – Retro pitch-shifter granular reverse delays

Eventide Anthology II Bundle

Eventide's great plug-ins have been around almost as long as Pro Tools. These are not just fun audio twisters but innovative audio processing tools that can help solve many audio problems and also twist the sound into something unrecognizable.

It includes the following:

- E-Channel – Configurable channel strip with gate, compressor/limiter with sidechain, and five-band 48-bit double precision parametric equalization

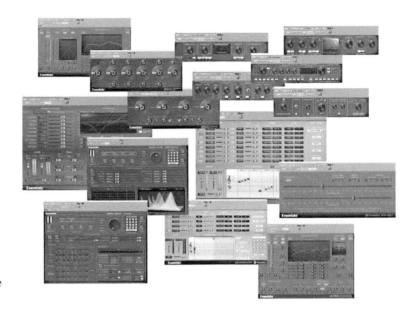

Figure 4.22 Eventide Anthology II Bundle.

- EQ45 Parametric Equalizer – Vintage 48-bit double precision four-band equalization with high and low cut 12dB/octave filters

- EQ65 Filter Set – Vintage 48-bit double precision high and low cut 18dB/octave filters, plus two-band reject or band pass filters

- Eventide Reverb – Nine room types from the H8000 with three-band parametric EQ (pre- and postreverberator), a compressor, and stereo delays

- H3000 Band Delay – Eight voices of delay with modulating filters; includes the Function Generator with 19 waveshapes

- H3000 Factory – Patch together any combination of 18 effects; includes the Function Generator with 19 waveshapes

- H910 – The original Harmonizer pitch shifter with delay

- H949 – Deglitched pitch shifting with delay and a randomizer

- Instant Phaser – Recreation of the world's first phaser

- Instant Flanger – Dedicated flanger with bounce

- Omnipressor – Dynamics processor with an attitude!

- Octavox – Eight-voice diatonic Harmonizer pitch shifter

- Precision Time Align – Track phase alignment tool

- Quadravox – Four-voice diatonic Harmonizer pitch shifter

- Ultra-Channel – Configurable channel strip with gate, compressor/limiter with sidechain, Omnipressor dynamics processor, five-band 48-bit double precision parametric equalization, micropitch shifting and stereo delays

ELS Vocoder – Orange Vocoder

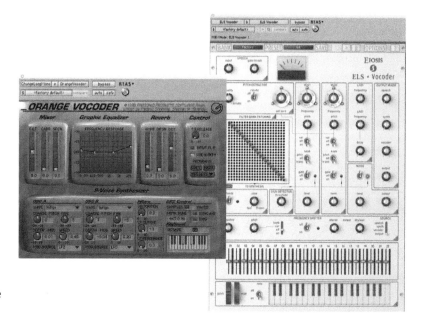

Figure 4.23 The ELS Vocoder and the Orange Vocoder by Prosonic.

The original Vocoders were analog synthesizer devices. The sounds from Moog and other synths were feed into the device that modulated the sound with a voice recording. The net result was a synth voice that was more or less understandable. These were popular in the 1970s for robot voices and other strange effects. These Vocoder plug-ins duplicate the function of the original analog devices in a simple to use Pro Tools plug-in. Although a Vocoder is strictly an electronic sounding voice or effect, the retro robot quality of the effect works in many science fiction or spy film projects. And who knows what strange sounds it can be twisted into making.

The Usra Major Space Station SST-282 by Princeton Digital

The Space Station is in orbit again. This fun hardware device from 1981 by Chris More is back and reworked into a TDM only plug-in by Princeton Digital. It is a delay-echo-reverb whose output delays are modulated to provide some wonderfully weird sounds.

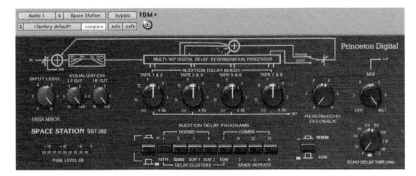

Figure 4.24 The Usra Major Space Station SST-282 by Princeton Digital.

4.7 Noise reduction and restoration

Often there is a need to remove hum, broadband noise, and many other unwanted sounds. It could be generator sounds, hum from lights or power lines or even the director or assistant director calling a direction to an actor.

There are also times when old sound files need to be restored for DVD rerelease or to use some great old sounds in a new film. There are many good tools to help the audio engineer–editor–mixer when these challenges come up.

CEDAR Tools

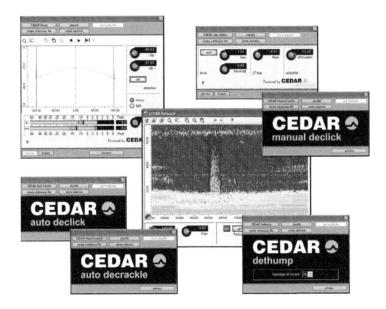

Figure 4.25 CEDAR Tools for noise reduction and restoration.

CEDAR Tools offer some great tools for restoring bad audio and removing unwanted sounds.

■ Retouch

A huge leap forward in audio processing, Retouch allows you to ident-ify and eliminate unwanted sounds as varied as coughs, record damage, squeaky chairs, and even car horns.

■ Auto Dehiss

Auto Dehiss embodies a more advanced algorithm than any previous dehisser, and it offers a unique "Auto" mode that enables it to determine the broadband noise content, removing this without the introduction of unwanted side-effects or artifacts.

■ Declip

A genuine declipping algorithm, Declip allows you to identify and remove most instances of clipping in a single pass. It provides two processing modes to ensure that you can remove problems without damaging the genuine signal.

■ Auto Declick

This plug-in automatically detects each click, removes it, and fills the gap with the best estimate of the material that would have existed had the click not occurred.

■ Manual Declick

Manual Declick is optimized for longer clicks and scratches, removing these without artifacts. Use this to eliminate extended clicks and pops; the maximum length handled is 4096 samples at 96 kHz.

■ Dethump

Dethump removes the low-frequency disturbances that you cannot restore using conventional declickers. The maximum thump length handled is 100,000 samples (a little over 1 s) at 96 kHz.

■ Decrackle

This automatically detects the tiny disturbances that comprise crackle and various types of buzz and distortion, removes them, and restores the audio with material indistinguishable from the undamaged original.

DINR

DINR is Digidesign's award winning broadband noise reduction plug-in. DINR samples a small section of the unwanted noise and then becomes a tunable

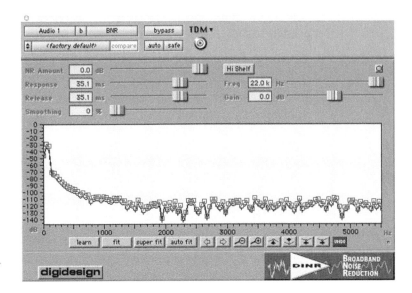

Figure 4.26 DINR by Digidesign.

noise gate that ignores the noise while gating on the signal. The result is good reduction of the sampled noise without any noticeable degradation of the original sound. There is an LE AudioSuite-only version as well, which is included in the DV Toolkit. This is a great tool for cleaning dialogue with unwanted sounds such as generator, refrigerator, air conditioner, and similar sounds.

Waves Z Noise

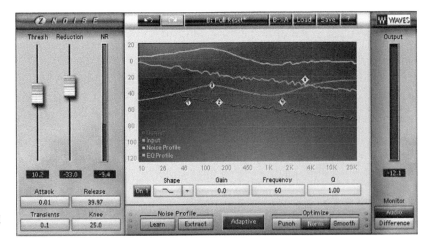

Figure 4.27 Waves Z Noise.

From hum to broadband noise, this may be the answer to your problem.

Features:

- Five-Band Noise Profile EQ

- Enhanced low-frequency resolution and time sensitivity

- Monitor the entire output or just the noise

- Exclusive Extract mode creates noise profile from sources containing signal and noise

- Adaptive dynamic detection for noise that changes over time

- Up to 24 bit 96 kHz resolution

- Mono and Stereo components

- RTAS and AudioSuite

reNOVAtor by Algorithmix

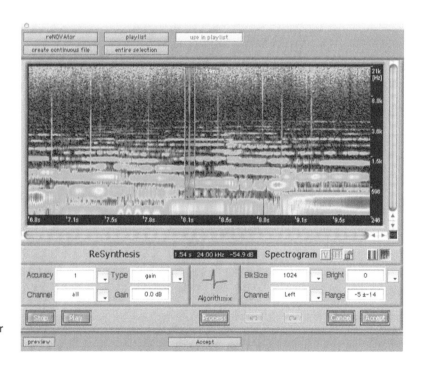

Figure 4.28 ReNOVAtor by ReSynthesis.

ReNOVAtor is used to find, identify, and very precisely remove unwanted sounds without messing up the audio you want to keep. ReNOVAtor does not leave holes in your soundtrack when removing the unwanted sounds. Instead, it "acts on an exactly tailored hole in the spectral representation of the processed signal that can be removed and replaced." In other words, it removes unwanted sounds, fills in the holes, and works very well.

Features:

- Flawless operation with up to 384 kHz sampling rate
- Audio fixes that were basically impossible until now
- No audible changes in desired signal and ambience
- Replacement of a spectral region (copy and paste)
- Automatic selection of harmonics belonging to a marked fundamental
- Automatic identification of clicks and spikes, tones, and harmonics
- Multiple undo functions
- All internal calculations in double floating-point accuracy (80 bits)

Waves Restoration Bundle

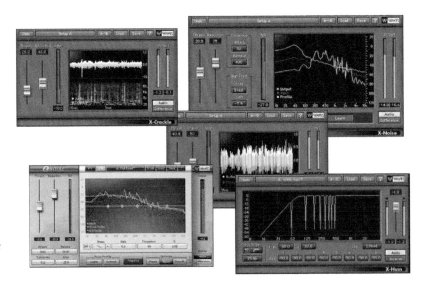

Figure 4.29 Waves Restoration Bundle. Five great noise reduction tools.

Waves Restoration bundle includes five great tools for removing unwanted noise from old recordings, vinyl records, old film sound tracks, field recordings, or whatever. With the release of many old films on DVD, television syndication, and forensics, these are a business in their own right.

It includes the following:

- X-Noise – Broadband noise reduction (Extra premo good)
- X-Click – Click removal
- X-Crackle – Crackle and surface noise removal
- X-Hum – Specialized ground-loop hum filtering
- Z-Noise – Adaptive dynamic detection for noise that changes over time

4.8 Utilities

Source Connect

Figure 4.30 Source Connect allows for the interconnection of two or more Pro Tools systems via the Internet.

Using Source Connect, it is possible to link several Pro Tools systems together via the Internet. One system can be used to control another with other systems "looking in." Any audio recorded on any system can be FTP'd or delivered via DigDelivery to the other systems. This is great for ADR sessions: the actor can be anywhere, even on set, and ADR can be recorded by the mixer in the postproduction facility. The director can even be in a third location. This saves time and money and frees the postproduction to be done remotely during production. It also saves travel time and expense.

The system can also be used to "look in" on the scoring session so that the producers and director can be in several locations at the same time. And this means several things can be done at the same time.

Virtual Katy

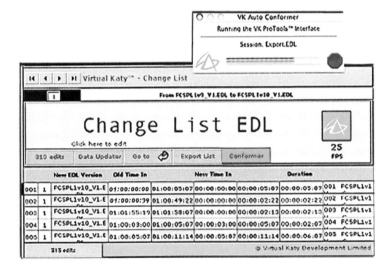

Figure 4.31 Virtual Katy conforms the audio edit to picture changes and monitors the progress of the editing.

The editors and software gurus on the *Lord of the Rings* trilogy needed a system to make picture changes after the edit was locked and so developed Virtual Katy. (Katy was the assistant sound editor who got stuck with making changes.) Change lists can be exported from Avid or Final Cut Pro and these lists are feed into Virtual Katy. Virtual Katy conforms the audio tracks to the new edit.

The way it functions is by taking the entire edit and shifting it forward in the time line by 1 h. It then moves the duplicates of the audio regions back to the original location while shifting by the changes.

Other functions included with the full Virtual Katy system include a system for conforming the original audio and a system for tracking the progress of the entire edit. The postproduction supervisor and/or the sound designer can track and preview the work as it is being done.

ADR Studio

ADR Studio by Gallery is used to cue and record ADR. The cue function creates printable cue sheets, while the plug-in imports this information into Pro Tools and inserts named regions in the cued tracks ready for ADR. It can also be used to cue and record Foley. It adds the cue beeps and even controls the record light. For more information on the use of this plug-in, see Chapter 5 "Dialogue Editing."

Plug-ins make every system unique and expand its usefulness. They customize the functionality to the specific needs of the user, and because they are managed by the iLok, they are portable to boot.

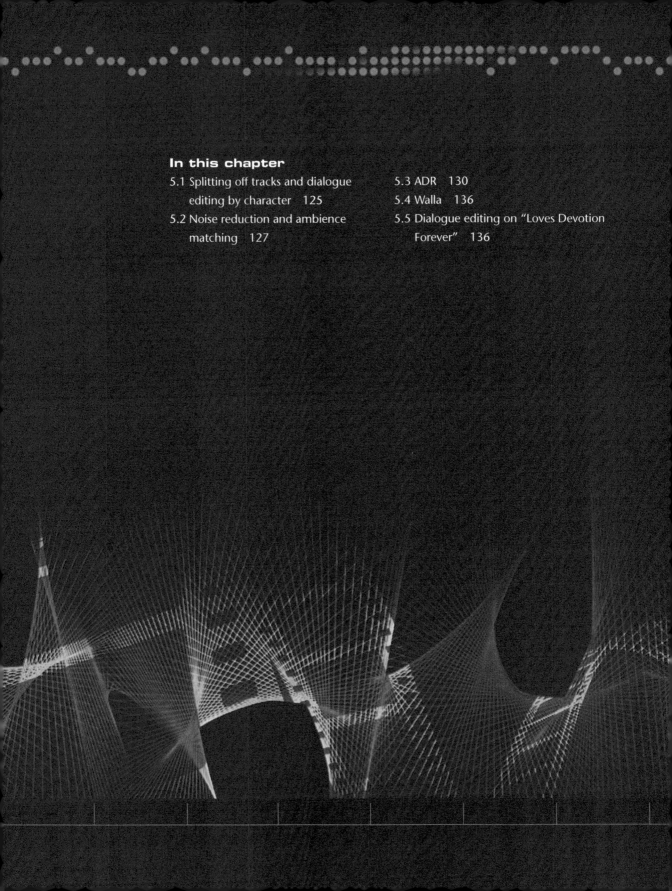

In this chapter

Dialogue editing and replacement

5.1 Splitting off tracks and dialogue editing by character

The dialogue edit will go much more smoothly if you start with well-organized tracks. The number of production tracks needed depends on the complexity of the project. Some rules to keep in mind: don't cut any sound clips up to each other unless they are derived from the same take. Although you can set a different level for each clip, in the final mix you want total control of the level of the individual clips, and this will be tricky if they are butted up together in the same track.

Every project will be given to the dialogue editor in a different track arrangement. The picture editor may have been very organized and placed all temp effects in dedicated tracks, temp music in other dedicated tracks, and the dialogue may even be more or less split off. Or everything may be a jumble with music, effects, and dialogue all over the place and even some dialogue out of sync. The project will probably be delivered as an open media framework (OMF) that will need to be translated into a Pro Tools project with DigiTranslator. The OMF may reference audio files that are delivered on CD or DVD or it may be self-contained, with all of the audio embedded into the OMF. Video may be delivered as a QuickTime or on videotape.

The OMF audio may be used in the final edit or the audio EDL used to recapture all of the production audio and replace the OMF audio one take at a time. Most often, the project will require a bit of both using any OMF audio that is clearly usable and replace any audio than is in any way questionable. On some projects all audio will need to be replaced simply for workflow reasons. (See Chapter 2 on workflow.)

When using BWF production audio the OMF can be used as the session audio files with the OMF simply linking all regions to the audio in the OMF.

Figure 5.1

When using the latest versions of Final Cut Pro and Avid, the BWF audio can be imported and used in the edit without transcoding or altering it. The OMF still contains this BWF audio and when it is used as the audio in the Pro Tools session, you are in fact using the original production audio.

All OMF audio needs to be time stamped before moving anything around. Even if the picture editor has already split off the production tracks before bringing the project into Pro Tools, it's still a good idea to "time stamp" all dialogue using the Time Stamp Selected command in the Region List pop-up menu.

The simplest approach to splitting off dialogue looks like an A B checkerboard layout with the first clip in track one, the second in two, third back in one, and so forth. The audio regions are extended and overlapped with quick fades to help hide the edits. This works very well and is the system of choice on many projects.

However, this layout can be confusing in mixing. Characters move from track to track and the mixer really keeps riding on the two production tracks. Many editors feel it is more logical to put each character in a dedicated track. This is how ADR (automatic dialogue replacement) is most often edited and it works when production recording is done with multitracks and multiple microphones. This is especially true with wireless mics; each person is already in their own track. However, production recording is different from ADR. ADR is all at the same basic level and it's okay to cut two takes together. Production tracks tend to be all over the place, so a cut in to a close-up or cut to the master butts two very different clips to each other in the same track, which is a problem in mixing. On a feature film, it is normal to have an A and B track for each main character. The character and supporting people may not have dialogue where two takes are ever butted together. Moreover, many will never be in scenes together and can therefore share a track. This will create scores of production tracks making mixing and track management complex and tying up all your resources trying to put noise gates and filters on all of these tracks. But this is how big projects are normally done. After all, they are big projects and should have big resources.

Production sound effects will need to be split off to several production FX tracks. These will be moved to the FX section of the mix and used with the edited effects and Foley to create a natural effects design. They need to be moved so that they end up in the FX stem after the mix. (See Chapter 3 on mixer layout and Chapter 8 about the dub.)

Every film and scene is different and different dialogue editors and mixers have different ways of working. Not all films need to have everything split off. One common strategy on smaller projects and documentaries is to limit the edit to three or four production tracks and try to group things into these tracks in the most logical manner and create a problem track; put all the noisy wide shots in the problem track (wide shots are always a problem); put shots that you think need special treatment such as noise reduction in this track.

5.2 Noise reduction and ambience matching

There are two strategies in editing around background noise in production tracks. Both work in some instances. The trick is knowing which to use when.

Dialogue editing technique one

The first technique is to fill in all holes with ambience. Carefully edit out any unwanted sounds like direction or a C stand falling over. Then overlap the cuts from one clip to another slightly, filling any holes left from cutting things out with ambience from this or other takes. Set up quick fades on the overlaps in an effort to make the ambience sound seamless.

This can produce a seamless ambience and works well on projects with a lot of dialogue and fairly clean tracks. It is often used on short films or simpler projects as well as almost all documentaries.

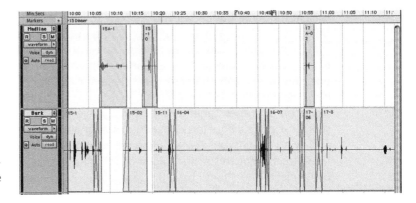

Figure 5.2 Editing dialogue by character while filling ambience holes.

Using this editing technique, the edit for the dinner scene in "Loves Devotion Forever" might look like this. (We will look at the actual dialogue editing of this scene later in this chapter.) All of Madeline's lines are in track one. Most of the movement and ambience is in the second track with Burk's lines. All unwanted sounds have been removed and all holes patched by moving audio around and copying and pasting. Short fades are used to smooth the edits and cover noticeable edits. The scene plays well and sounds good. (This example is included on the Website.) It is tricky to mix because some of Burk's lines are rather different from each other, some are not good at all, and some are butted together in the same track.

If the background noise is shifting dramatically on the edits and some of the dialogue will need to be replaced, this technique may not produce usable results. Moreover, if the mix is intended to produce an M and E for foreign language replacement, ambience tracks will be included in the effects editing, producing even more background. Therefore, another technique is often used on features.

Dialogue editing technique two

In this case, the dialogue is cut as tight as possible to the modulation stripping out all sound between the lines. Short fades can be used to ease in and out of lines where the background can be heard "popping in" with the lines. The background from a clean section or sections of audio is then used to create a production sound background. Usually, the ambience and/or background will run through the entire scene and will be part of the effects tracks. A good ambience will fill all holes and help establish the scene.

With noisy production audio that will not be replaced, the ambience needs to match the background of the production audio. Often the production mixer will record a wild track of ambience that can be used for this track, but often the clean noise between lines work as well or even better. At times, the track will be rather noisy, but necessarily so. It needs to match the background noise of the dialogue. At times, a truly horrid mix of camera noise and rumble must be used to "save" the scene One old-time editor referred to this as "dragging a dead mule through the scene" (colorful but apropos). It was something that no one wanted, but it was the only way to make the production audio work. Often the "dead mule" was camera noise from the same noisy camera.

Figure 5.3 Dialogue editing technique two creates a seamless background and movement track from the production audio (track four). Although this is the same scene as in Fig. 5.2, in this case Burk's and Madeline's lines are still in tracks one and two, only now they are cut tight to the modulation so that there is no sound between lines, all movement and background sound now sounding on track four. One line of problem dialogue for Burk is in track three. This line needed clean up and so was placed in a separate Burk track. Often this type of splitting is done for every other line, creating sort of a checkerboard that allows for adjustments to each line. In this case, only the problem track is split, and was eventually replaced with ADR.

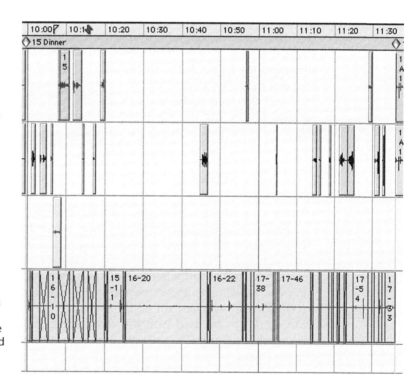

There is a great trick on the TDM HD Pro Tools systems called voice borrowing. On these systems you can assign a "voice" to a track. Normally each track is set to dynamic (dyn) and has its own voice, but on TDM systems you can assign a voice to several tracks. Only one track on a voice can play at the same time, that being the lower track number; in other words, the track nearer the top of the Edit window. Therefore, if you lay an ambience in on track two and tight cut dialogue in on track one, with both tracks assigned to the same voice every time, a line plays the ambience drops out. Neat trick in some cases. The Auto Region Fade preference in Operations sets the speed of the switch from 0 to 10 ms, so you can make the switch a very fast fade. Actually all regions now fade in and out to this setting. Also, when the dialogue tracks are made inactive for exporting the FX stem, the ambience now runs through the entire scene without switching.

In either case, the AudioSuite plug-ins can be used to clean up and match the dialog on a clip-by-clip basis. Select the clip you want to work on in the timeline and open one of the AudioSuite plug-ins. Gain can be used to bring up very low lines. DINR (digidesign intelligent noise reduction) can be used to clean up some if not most of the background noise. Sample a small section of background sound only and let DINR use it as a guide to remove all similar sound. Four band equalizer (EQ) can be used to remove rumble and help to match microphone perspective from take to take. All of this is done to the clip only, so final adjustments will always be made in the mix where all the tracks are heard at the same time. Don't overdo things here. Always do the least necessary to get close to where you need to be. AudioSuite processing is not destructive; the originals are still in the region bin. If you do go too far with processing, the mixer will want you to put the original back in the edit.

On the Website

These two editing techniques can be experimented with the website project. The scene is the dinner scene, scene 15, memory location 12. Experiment with the two to get a sense of the appropriate technique for each situation. For experimenting with voice borrowing, on HD systems, set all four tracks to the same voice and compare this technique to keeping the movement and background track sounding through the entire scene. Solo the movement and background track to see how this production track can be used with the other effects and music tracks to create a complete M and E (music and effects) track for foreign language voice replacement.

5.3 ADR

We are now ready to cue or "spot" the ADR. For years, this was done with a process called "looping," where the audio clips were cut into loops with cue pops and played in sync with a film recorder. The looped guide track served as a sync guide for the actor who rerecorded the lines in a sound booth or small stage.

Figure 5.4 ADR is usually recorded in a voice booth or on a larger stage with video playback.

ADR replaced looping when automated dialogue replacement systems were developed. ADR historically was recorded in a studio equipped with an ADR system. In ADR the entire film or reel is played as cue pops are generated by a computer. The actor hears three cue pops about 1/2 second apart just before the line, and the line to be replaced, and sees the picture as they rerecord the line. Actors can get quite good at matching sync; however, some may like a different cueing system. Some don't like the pops and prefer to cue to their movement on screen, others may want only two pops, faster pops, slower pops, and so forth. Many ADR systems let you customize to the actors, wishes, whereas others don't.

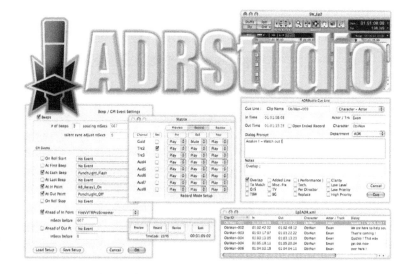

Figure 5.5 ADR Studio from Gallery allows for cueing, printing cue sheets and can be imported directly into the ADR session. It creates pops, lights the record light, and interfaces with both Pro Tools and Virtual VTR.

Cueing or spotting the ADR is an art in itself. The person doing the cueing must decide which lines can be used and which ones need to be replaced. What problems can be fixed in the mix, what problems cannot? If there is a major problem of noise over dialogue, then the whole line is replaced with ADR. And if there is no way to match the ADR then the whole scene will need to be replaced, or at least this character's lines. The performance is also an issue. Sometimes the lines are replaced in an attempt to improve a weak performance. In this case the director and/or the producer need to be part of the process. Often, new off-camera lines are also added.

Cueing should take place in a pseudo-mixing room or even right on a dub stage. The person doing the cueing needs to be able to hear what is on the track in a critical listening environment. It's also nice to have some basic mixing tools and noise reduction equipment like the tools that will be used in the mix to determine if a problem can be fixed. There is no attempt to actually fix any problems here, only decide what is fixable. Cueing ADR is a diagnostic process.

The lines to be replaced are listed by actor and timecode and given a memory location in the dialogue session. The exact line as used in the film, not just from the script, is transcribed. The cue sheets need to be copied and given to the effects editor, who is also cueing, as the ADR is being cued. The effects editor will need to know which production tracks are going to be removed so they can cue proper effects and Foley in these areas.

In the ADR session the Director, ADR mixer, ADR editor, and actor are present. It is important for the actors to put themselves into the scene. If they are scream-ing over a jet engine as fire burns all around, it may be harder for the actor in

Figure 5.6 The ADR Mixer uses mic placement and minimal EQ to match the sound of the original production recording.

ADR to recreate the performance from the set. They are warm and comfortable in a studio, not on an aircraft carrier. Sometimes actors need to run outside and run around the parking lot to get winded and into the scene – whatever it takes.

The ADR mixer has a difficult job. They need to record the line cleanly, match the sound of the nonreplaced lines, and keep the sound from sounding like it was recorded in the studio. Anyone can record clean audio, but it takes an artist to record the not-so-perfect, nevertheless just right, audio. Microphone selection and placement are the strongest tools of the ADR mixer. Some EQ can be used, but it's best left to the dub later on. Many ADR mixers use two mics, one of which is the same model used in production. One mic is placed some-what close (never right on mic), the "production" mic placed at a distance.

Usually two or three good takes are saved. This way the ADR editor has some material to tweak. The ADR edit is all about using the right words from the right takes. Tweak until it fits. Everyone has noticed bad ADR in a movie. No one notices good ADR. The tracks are usually cut so that each character has their own track in the scene.

Recording ADR with ProTools

Figure 5.7 Studio mics are often used for voice over and to reinforce production mics typically used in ADR. Ideally the same mic used in production should be used in ADR; however, many ADR mixers also use a second mic.

The ADR should be cued and transcribed by the ADR cueing assistant or dialogue editing assistant. The cue sheets need to show the session timecode of first modulation of the line to be replaced. The exact line is transcribed into the ADR script. Keep one copy of the script on loose single-sided pages. In the studio you will want a wide music stand to hold the script so that the loose pages can be laid out with no need of turning or handling.

Although ADR has been recorded with ADR computers, and still in some studios, ADR is often recorded directly into Pro Tools. This requires a good studio space or voice booth. Avoid a booth that sounds like a booth; a more open studio or even a live room may sound better. Work with the mic positions to match the production sound as closely as possible without using any reverb and minimum equalization.

As we are not using an ADR computer, you will need cue pops. These can be cut from 1000 Hz tone generated with the tone AudioSuite plug-in. Edit 3 one frame pops about 20 frames apart and bounce these to disc. You will also need to set up a video monitor and headphones for the actor in the studio space.

Create several new tracks and name them ADR 1, 2, and so forth. Check the input assigns and set them to the mic input(s) that you are using for recording. Also create a new track and name it *cue*. Import and drop the three cue pops in the cue track at a point just at the start of the line you want to replace. You can copy the cue pops and place a set before all the lines you need to replace or simply move them up as needed.

Go to the cued timecode and select the production track region with link timeline and edit selection on. Turn Link Timeline and Edit Selection off. Solo the cue track and the production track containing the original line. Set the in and out brackets in the timeline to a point just at the start of the cue pops and the end of the line to be replaced. Set loop playback in the operations menu. Record enable ADR 1.

Begin playback. The actor will hear the loop playing over and over in their headphones as they try to match sync. When you feel the actor is ready, set the transport control to record and click play. This will record one take and stop. Check the line by soloing it on the Pro Tools mixer and unsoloing the cue and guide tracks. If you like the line rename it in the region bin. If you want another, record enable ADR 2 and unenable ADR 1. If you don't want to keep the take, select it in the region bin and from the menu at the top of the region bin select delete. If the actor is only able to hit sync on the third or fourth loop playback, you can set loop record and record several takes as one master region with several subregions of each take.

ADR editing

Once the ADR is recorded it needs to be tightened up and arranged by the ADR editor. The ADR tracks as recorded may have all characters in the same three or four ADR tracks with several takes for each line in the tracks. Each main character should have their own ADR track and other characters can share ADR tracks as long as they are not in the same scene together. The best take from the ADR session is placed into the proper track and aligned for best sync. This may involve cutting and pasting the line around or even cutting several takes together to achieve perfect sync and performance.

A very useful tool for aligning ADR is VocALign from Synchro Arts. The original guide track is placed in one track of the plug-in and the recorded line in the dub track of the plug-in. A volume plot is generated from each line and compared. The editor has some control over the alignment of the two plots. When the editor is happy with the alignment, time compression and expansion is used to match the sync of the dub track to the original. This is so good at pulling the sync together, that it is possible to record the dub lines wild with no picture or guide track.

There are two versions of the software, VocALign project and VocALign Pro that allows the editor to match "sync points." Like all time alignment tools, overdoing the compression or expansion will create very undesirable artifacts. What this means, though, is that the editor can now use the performance the director likes without worrying about small sync problems. In fact, the alignment can be done right in the ADR session with the director present to accept or reject the take without wondering if the editor can "fix" the sync in editing.

If the production audio is a noisy, VocALign may not be able to "read" the volume plot making it impossible to align the ADR to the production. However, if there is one good sync take, it can be used to align the best performance take.

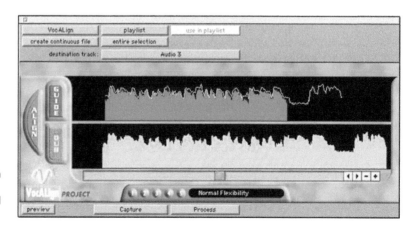

Figure 5.8 VocALign can be used to pull the ADR into sync with the original production audio.

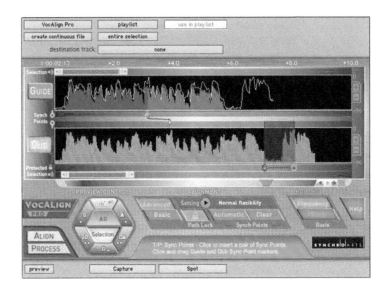

Figure 5.9 VocALign Pro offers more control and better performance over the basic VocALign.

5.4 Walla

One special type of background is *walla*, a group of human voices blending into a background. Editors, distributors, producers, and directors often disagree on whether walla is an effect or dialogue. Should it be cut by the effects editor and/ or end up in the effects stem? Walla is often recorded by the ADR editor who schedules a walla session with a group of walla actors. (yes, there is such a thing) This walla is cut into an ADR track. Or, often the effects editors put together their own walla session or pull walla from a library. In this case the walla ends up in an effects track and possibly the effects stem. Whichever system you use, never premix the walla into the background; you may need to record it into the dialogue stem in the dub. Foreign language dubs may require foreign language walla. If not, it is a sound effect and should be edited into an effects track.

As a general rule, completely nondescript walla voices should be edited into an effects track. Any walla where the actual words can be understood, or even a sense of the language, should be edited into a dialogue track so that it can be replaced in any foreign language dub.

5.5 Dialogue editing on "Loves Devotion Forever"

The dialogue edit

The OMF was imported with the DigiTranslator option and the Pro Tools edit created.

On the Website – The Dialogue Edit

The website includes several Pro Tools sessions from "Loves Devotion Forever." The first is named as OMF Import. This is the Pro Tools session derived from the exported OMF audio exported from Final Cut Pro. This could also have been exported from Avid or any number of picture editing softwares. Most of the dialogue to be used in the dialogue edit is in tracks one and two. The other tracks are mostly temp music and some temp effects and temp ADR. These sessions can be opened and edited on any Pro Tools V7 or V8 system, LE, M-Powered, or HD. The finished dialogue edit is named DIA edit and contains all the split off tracks as well as the temp ADR. Memory locations have been created for each scene, as well as show hide presets for dialogue, temp tracks, and production effects.

The other two sessions are edit type one and two showing the two systems for splitting dialogue in the dinner scene, memory location 12.

You can follow along with the edit and try your own systems of splitting off the dialogue.

Figure 5.10 The working title of the film was *Blackheart*. The film was edited on Final Cut Pro with cinema tools managing the media database. Several files were exported from Final Cut Pro, two of which are the basis of the Pro Tools session: Blackheart_Final [red carpet].omf and the QuickTime, Blackheart_ Final copy. The other files are the Final Cut Pro project, the Cinema Tools database, the cut list for the negative editor, an archive project in XML, a picture EDL (edit decision list), and an audio EDL.

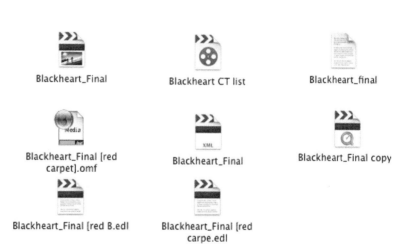

Blackheart_Final

Blackheart CT list

Blackheart_final

Blackheart_Final [red carpet].omf

Blackheart_Final

Blackheart_Final copy

Blackheart_Final [red B.edl

Blackheart_Final [red carpe.edl

Once the initial session was created from the OMF, the QuickTime picture and audio guide track were imported and the two pop checked for sync on the countdown leader.

Notice that the production is doubled in tracks one and two in many places. The production audio was recorded into two tracks whenever only one

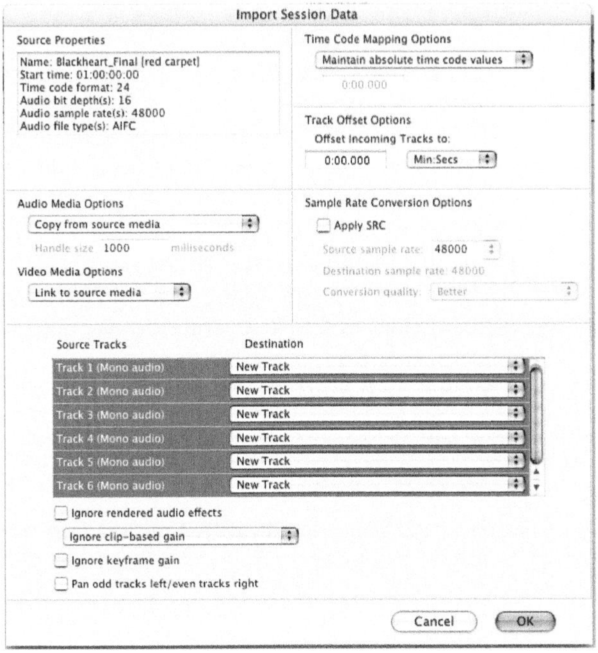

Figure 5.11
DigiTranslator was used to import the OMF into Pro Tools and create the initial dialogue session.

Figure 5.12 The initial import of the OMF showed disorganized tracks, the temp audio was in all tracks, even audio one. This is not uncommon, and it is not a problem. The first step in editing will be to time stamp and organize the tracks to begin production dialogue editing. New tracks were added and tracks named for the dialogue edit. This has already been done in Fig. 5.10. Initially two tracks were created for Madeline, one for Burk, two for other characters, two for temp ADR and effects, and four for temp music.

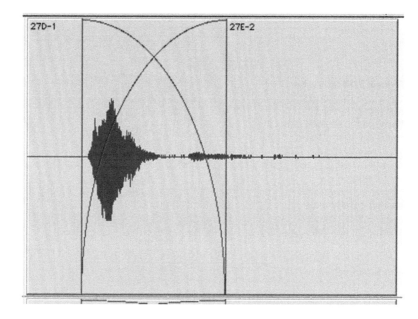

Figure 5.13 The OMF had also imported the automated fades into the Pro Tools edit from Final Cut Pro. These were also present in the fade files folder in the Pro Tools Session folder.

Figure 5.14 All temp music was moved into the four temp music tracks. The temp music tracks and the guide track were all made inactive and set to hide in the show/ hide menu. In the event they were needed in the future, they could be made active.

microphone was used on set. When the two tracks are at the same level, one is simply deleted. If one of the tracks is a safety track at −10 dB, unless it is needed it can be removed or moved to an inactive track on the chance it may be needed in the future.

All production regions were time stamped in their current location using the Time Stamp Selected – User Time Stamp before any moving of these regions.

Figure 5.15 All scenes were given location markers at the first frame of the scene. From the memory locations window any scene could now be selected by clicking on the corresponding memory location. Show-hide state locations were also created for the dialogue tracks and the temp tracks. Memory locations and show-hide states are created in the new memory loc pop-up menu in the memory location window (click on the Name button).

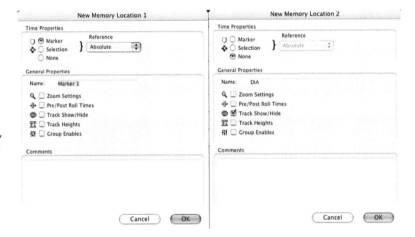

Figure 5.16 The new memory location menu is used to create not only markers at certain locations, (setting on the left) but can be used to create show-hide states, zoom settings, group enables, and other presets. They can also be used to cue ADR, Foley, effects, and other cues.

In Figure 5.18, the dinner scene in "Loves Devotion Forever," initially the imported picture edit had two tracks of production audio containing a mixture of types of tracks. Some were duplicate mono tracks, most were two mic setups, one wireless, and one boom. The audio is not good in some places, and will be a challenge to work with. Some may need to be replaced. A visual examination

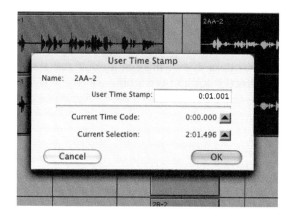

Figure 5.17 Before any dialogue was moved, it was time stamped with the Time Stamp command in the Region List Pop-up menu. The current selection was used.

Figure 5.18 The original production tracks and picture for scene 15, memory location 12, the dinner scene. The three windows printed on the video below the image show: Upper left – Production audio timecode Lower left – Video timecode for this frame of video Lower right – Film keycode number for this frame on the original 35mm negative. Film stocks are flashed with numbers and machine readable barcodes that show up after the film is developed.

of the tracks revels that the first region is a duplicate track in one and two. The next two regions are two mic setups, different audio in tracks one and two. The next two regions are duplicates again and the last one is another two mic setup.

The tight cutting technique we looked at before in Fig. 5.3 required that all lines be cut as close to the dialogue as possible. When using this technique, remove anything that isn't dialogue. No ambience between lines, nothing between the lines. Then find a good ambience quite similar to the original and lay it in on an effects track. If necessary, set up quick fades on the production lines so they don't pop on or off. In this example, the production movement and ambience have been moved to track four and all holes patched with ambience and movement from other takes. This creates a seamless background that can be placed in the effects tracks and mixed into the effects stem. Normally this would not be a continuous track with movement and effects as it is in this scene, but, rather, the effects moved into a production effects track and the ambience would be placed in another track.

Figure 5.19 In this example one of Burk's lines has been split off into a new Burk B track. His first two lines at the beginning of the scene on the porch are weak and noisy. The third line does not match the sound quality of the other two lines, which will need special treatment or possibly replacement. Mixing the production lines would be difficult or even impossible if they remained butted up to the third line. (As it turned out, all three lines were replaced with excellent results.)

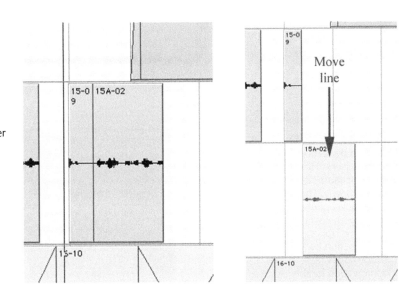

Because this scene has long sections where there is no dialogue, only movement, it was possible to create this continuous background with movement. With the addition of a little Foley and one or two tighter effects, the effects stem will be complete. The scene will play with the dialogue out. If the scene does need ADR, or when it is "dubbed" into another language, it is ready to mix. If it were cut using the other "hole patching" technique, it would need major reworking.

Figure 5.20 When tight cutting dialogue the strip silence function in the edit menu (window menu in V6) can be used to remove the audio between lines. It tends to clip off the tail of the line, and usually requires tweaking or trimming.

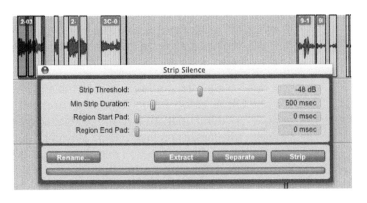

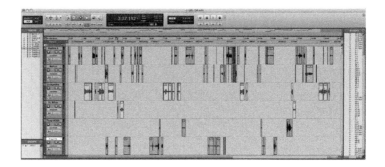

Figure 5.21 The final dialogue edit with two tracks of production effects. This tight cutting technique only work swell when the dialogue is fairly clean of background noise. Audio with a lot of background noise requires the patching technique.

Initially there were four dialogue tracks laid out: Madeline, Burk, DJ/Agent/Newsman (who were all sharing the track as they never appeared in the same scene), and the fanatics. Later, this was expanded to include a Madeline B track and a Burk B track as both had lines that needed to be split. Effects were split off to a production FX 1 and FX 2. All the temp ADR consisting of added lines was moved into a temp ADR track to be replaced later with the actual ADR. (The dialogue edit is included in the project Web site.)

ADR and walla on "Loves Devotion Forever"

Originally it was thought the only ADR needed was the added lines from Burk, Madeline, and the Agent on the answering machine, the premiere screening announcer, and Burk's and Madeline's lines as they run out of the theater at the premiere. After attempting to "save" Burk's dialogue on the porch before scene 15, the dinner scene, it was decided to replace these lines instead of trying to "fix it in the mix."

Because Madeline's lines butt up to both the Agent and Burk on the answering machine, two ADR tracks were created, ADR Burk/Agent/Announcer and ADR Madeline. VocALign was used to tighten Burk's lines on the porch with good effect.

A walla session was put together to record walla for the theater and the final scene at the pier. Party walla was pulled from the effects library. About 30 "Brookies" (Brooks Institute film students) showed up and were recorded in the Institute screening room.

Some of the audio regions needed to be cleaned up. Some had a trace of unwanted noise; some had sounds that needed to be cut out. Often some small sound happens right in the middle of dialogue and yet can still be removed. Hopefully the sound is not right on a word so that it can be edited out and the hole filled.

Some lines were processed with AudioSuite plug-ins. Some needed gain applied, some broadband noise reduction, and some clip removal.

143

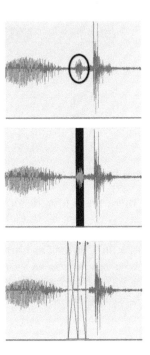

Figure 5.22 In this example, a wireless mic was bumped during the take, seen in the circled area in the upper image. Although the sound is small, it is more than can be repaired with the pencil tool drawing in a new waveform. Therefore, in the second image the unwanted sound is selected and was deleted. Then a short section of audio from just before the hole was selected, copied, and pasted into the hole. Very short fades were used to smooth the edit as seen in the bottom image.

A great trick when using AudioSuite processing is to make a duplicate track before doing any AudioSuite processing and then hide and make the new track inactive. This way if you do need to return to the original you can bring the track back and move regions as necessary. Once the plug-in is used, a new usage of the audio is placed into the track. And while the plug-in is not destructive, the original media is in the region bin with hundreds of other sounds. It can take several minutes to find the original media if it becomes necessary to remove the plug-in. If there is a duplicate track, which is kept inactive, the original media is always right there under the filtered sound, ready to be moved back to the original track and spotted into position in seconds.

Figure 5.23 Several of the "fine" lines in the front yard had a tiny amount of clipping. There was no need to ADR for this, the problem was small and the Bomb Factory clip remover improved the sound.

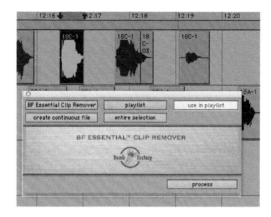

Figure 5.24 The apply gain plug-in was used on several lines that were just low enough to be a problem in the mix. No problem – applying gain brought them up to a better level with little extra noise.

Often the problem is noise. Hum can be filtered with a good parametric filter, but broadband noise like an air conditioner or a furnace is a problem. These kind of sounds can be helped with broadband noise reduction (BNR). This plug-in samples the noise and then creates a tuned noise gate to remove that characteristic noise.

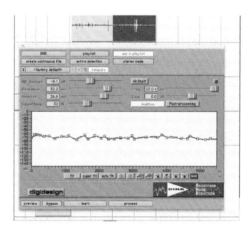

Figure 5.25 The BNR AudioSuite plug-in is used to remove noise that is at the same frequency as the human voice. It samples the noise and then removes that sound without affecting the dialogue.

Hum and sounds that are above the range of the human voice as well as below the human voice can be reduced with a good notch filter or "parametric" filter. These filters can be tuned to a certain frequency that is then attenuated. There are many such filters available as plug-ins. The seven-band EQ that comes with all Pro Tools systems is a good tunable filter able to remove hum, hiss, and rumble.

One slick system for removing hum is to turn the gain on one of the bands all the way up, the Q to the narrowest peak possible. Then tune the frequency

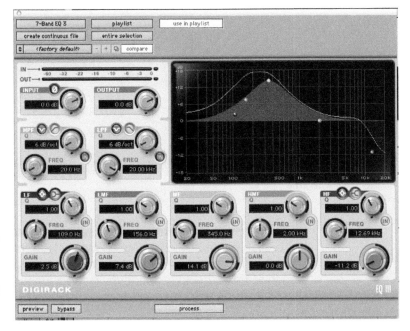

Figure 5.26 The seven-band parametric by Digidesign can be used to remove hiss, hum, and rumble. The FREQ knob is adjusted to the frequency of the sound or hum to be removed, the GAIN knob is turned down, and the Q is used to set the scope of the filter.

until the hum gets very loud. Then turn the gain all the way down. It is easier to hear the hum get louder than softer. Rumble and hiss can often be removed with just the HPF or high pass filter for rumble and the LPF or low pass filter for hiss.

With all of the dialogue split off into tracks and cleaned up, the dialogue is ready for mixing. Part of the key to good audio cleanup is to understand what can be done and will be done in the mix. You don't need to do everything here; in fact you want to do as little as is needed.

Dialogue is the heart of the soundtrack. Although the big effects and dramatic music get the attention, it's the dialogue that must be clean, understandable, and "just right" for the scene. People often talk about the old silent movies. These were projected with live music and often basic live effects. But without dialogue, the film is silent.

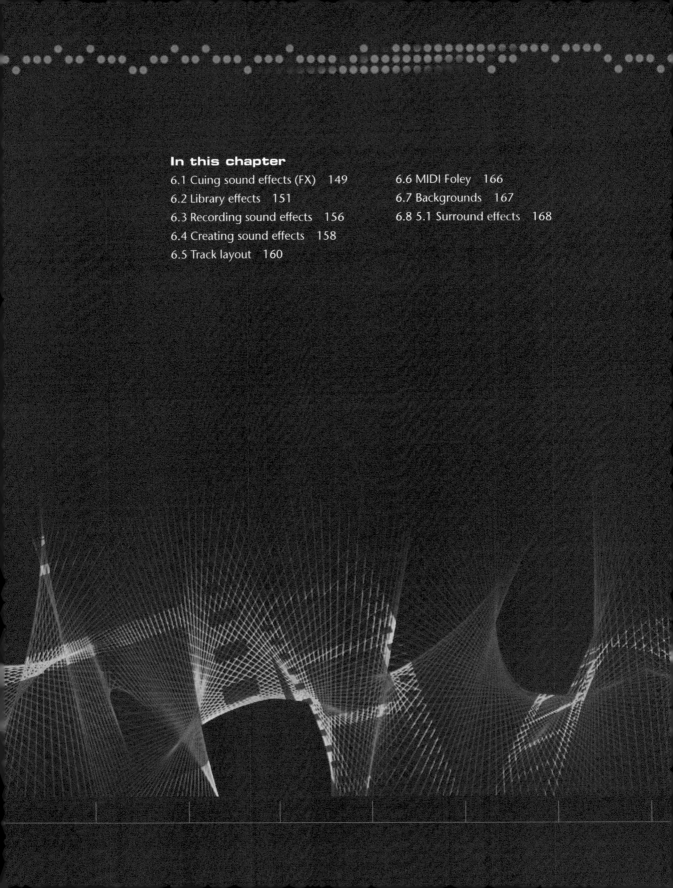

In this chapter

Sound design

6.1 Cuing sound effects (FX)

Sound effects are cued differently than automatic dialogue replacement (ADR). Critical listening is not required; critical thinking is. The FX editor needs to previsualize the mix, also called the *dub*. What is needed to make this scene play? Just because there is a production track doesn't mean the effects are there or clean enough. Moreover, the dialogue editor may have cut them out. In this case you may want to cut the effects from the production track into your effects track. Or the dialogue editor may have moved all of the effects into one effects track for you.

You may need to reinforce certain sounds, a doorknob, a match lighting, whatever. There will be sounds missing because they were never recorded or are sounds for visual effects. Subdivide the sounds into components. A massive chandelier crashing through a heavy table is not a sound you will find in a library. Nor do you want to attempt to record it! And the production sound, if any, is no doubt the sound of a plastic chandler smashing through a breakaway table. Look for a heavy wood crash, a heavy iron hit of some kind. What else can be seen? Chains rattling, wood splintering, glass breaking, plaster falling around the chandelier – you may need 10 tracks, but you can build the sound from components.

Make a list of the sounds you need and their timecodes. Try to remember the look and feel of the scene. Start listening to library effects. When you find something don't just dump it to your drive, write down where it came from on your cue sheets. Get several of everything. Many sounds that sound perfect now may not fit with the picture when you try it. You will also need to go out and record many of the things you need. The real secret to effects editing is finding or recording the perfect sound. Don't think too much about the real sound, think about the scene and the film as a whole. The right sound may not be the most "real" sound, but the perfect sound.

See an animal, hear an animal – the art of sound design

There is an old "rule" in sound effect says that everything seen on camera must make a sound. Paintings on the wall may be quiet, while in some films, for example Harry Potter, even these make sounds. Animals are almost always heard, even the most stoic and quiet species tend to be chatterboxes on camera. A bear will always growl, a lion will always roar, and even a deer may whimper, this despite the fact that a deer doesn't whimper. If the animal can't make sound on camera because it isn't actually doing anything, it will do it off camera.

Although this is an old rule, and most editors would say a stupid rule, there is a logic behind it that must be understood. It is the goal of the film to tell a story. Everything in the film including the sound effects must be true to the story. Reality is only important when it is an element in the story. In "2001 A Space Odyssey," the only sounds heard in the vacuum of space are breathing sounds inside the spacesuits. This is the intent of the filmmaker; therefore, it is necessary in the sound design and is critical to the story. The tag line for Alien warned "in space no one can hear you scream." But when watching the film that's about the only thing you can't hear. Every space ship rumbles, every huge latch makes a resounding metal bang as it opens, and even radio antennas produce a beeping, crackling sound. The Alien snarls and growls and computer screens beep and make Teletype sounds. All of which is true to the story and appropriate to the filmmaker's vision.

If a lion or a bear or even a snake needs a little help from the SD FX to seem more threatening, so be it. If a deer needs to whine like a dog to feel sad, and the story is supported by something that pedestrian, go for it. A producer once demanded that the effects editors take out the sound of a raccoon and use the sound of a spider monkey saying "That's the sound a raccoon would make if it had a choice." Keep in mind that this was in a film with a talking bear. On-screen reality is relative to the story.

Many of the sounds used in the FX tracks will not be intended to provide the sound of any real or unseen thing in the scene, but are purely an effect – a drone, whispering voices, a heartbeat, or whatever. These sounds are often referred to as nondiegetic sound as is the musical score. Diegesis is a Greek word for "recounted story." Diegetic sound is actual sound from a source on camera or from off camera. Nondiegetic sounds help to create the feel and mood. They may even take the place of or augment a sync effect. A light falling to the ground may have a certain diegetic sound, but the effects editor may instead use the sound of a piano falling over to create a startling and unreal effect. At times such sounds may be much more like a musical score; they may be very dramatic or very subtle.

Effects editing is very creative and very technical. The sound must be high quality, at times challenging the capabilities of the most advanced recording and playback systems. It must complete the soundscape of the production recording and it must be imaginative, artistic, and emotional.

There is a tendency to throw in the kitchen sink in the effects edit. If too much is put into the track it turns to mud. The effects editor and especially the effects mixer in the final dub need to know how to feature sounds. When the director and cinematographer figure out the framing of the shot, they feature the things in the frame and remove everything else. Even a wide shot of Hong Kong is not a shot of 10 million people doing 10 million things; it is a shot of one thing – Hong Kong.

The featured sound is most likely to be the dialogue when the actors are speaking. Nothing in the scene should compete with that unless the dialogue is not the featured sound for some story reason. In Close Encounters of the Third Kind the dialogue in the opening scene is buried under the wind of the sandstorm. The wind is the featured sound until the airplane engines start. Then the engines are the featured sound. At the end of the scene one line of dialogue gets to be featured. The featured sound is certainly not the only sound, but everything in the scene is supporting the featured sound. This may involve finding effects that do not compete with the featured sound. If a jazz band is playing, a saxophone will compete with the dialogue because it is a similar sound. If the band is a piano trio, they can be mixed more forward under dialogue without competing.

The same is true of effects. Pick sounds that don't compete with the featured sounds. If a train is passing by, it's likely that it will blow its horn (see an animal, hear an animal). But it should not blow its horn until it can be featured; don't compete with the dialogue. The same is true of music; at times the score will be the featured sound. If the plan is to have the score carry the chase scene, then a massive effects edit is not only pointless, it's counter-productive. If the effects are mixed up, they will only compete with the music. The music may be scored to let the effects be featured at times in the scene. If there is a big crash or explosion in the scene this effect will most certainly be played up, and the music should be scored and/or mixed to let this be featured. If not, the mix will be mud.

Many of the featured sounds are integral to the story and are in the script. The actors may react to them, but in many cases they are in the script without any outward signs in the action. The effects editor needs to have the script to understand these featured sounds.

6.2 Library effects

Many of the sounds needed in the edit may be found in an effects library. There are many good libraries on the market. Sound Ideas, Hollywood Edge, Network and scores of others. The libraries are usually sold as CDs, but some effects can be downloaded online. The CD option is usually much cheaper per effect, but you may buy 5000 effects and use two. In the long run it's still a good investment. They are also usually in Audio Interchange File Format (AIFF) format but are becoming available as broadcast wave format (BWF). The entire library can be placed on a drive and imported as needed. The rights are normally "buy out." You can use them without paying royalties for as long as you own the CDs, but you can't resell the effects.

Libraries contain thousands of sounds and may have the sound you are looking for if you can find it. An effects library without a great search engine is useless. If you are looking for a waterfall, a CD named "water effects" with 100 effects called track 1, 2, and so forth is a waste of time so profound you might as well just fly to Niagra and record the effect. It will be faster. A good search engine hits on words like "wood," "hit," "fall," "break," and "by." Search for expressions like "car by," not "car passing," and "body fall" not "person falling down." Most of the good libraries have their search engine online. This helps sell the effects libraries, but it is a great advantage to the editor. You can search from anywhere you can get online.

Let's say we are looking for wood crashes for the chandelier crashing through the wood table effect we discussed earlier. Let's go to Sound Ideas and see what comes up.

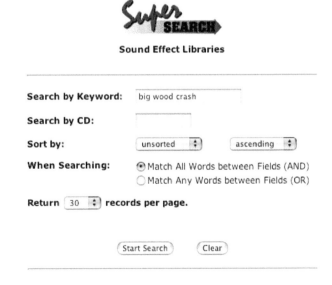

Figure 6.1 Sound Ideas online search engine. Sound Ideas is one of the largest suppliers of sound effects in several formats including MIDI.

We have entered "big wood crash." This is only one of many sounds we will need to make this effect, but it's a good start.

Search: Results

Displaying records 1 thru 7 of 7 records found. (7 records displayed).
Sorted by field: unsorted

Search Again

CD	Track	Index	Description	Time	Product	UR
PS01	30	1	CRASH, VARIOUS BIG METAL, GLASS AND WOOD CRASHES	0:18	PowerSurge 1 SFX Library	URl
PS01	31	1	CRASH, VARIOUS BIG METAL, GLASS AND WOOD CRASHES	0:18	PowerSurge 1 SFX Library	URl
PS01	32	1	CRASH, VARIOUS BIG METAL, GLASS AND WOOD CRASHES	0:19	PowerSurge 1 SFX Library	URl
PS01	34	4	CRASH, WOOD BIG WOOD CRASH WITH HEAVY IMPACT	0:04	PowerSurge 1 SFX Library	URl
PS01	34	5	CRASH, WOOD BIG WOOD CRASH AND SPLINTER	0:05	PowerSurge 1 SFX Library	URl
PS01	34	6	CRASH, WOOD BIG WOOD CRASH AND SPLINTER	0:09	PowerSurge 1 SFX Library	URl
XV04	99	3	CRASH, WOOD BIG WOOD CRASH AND SPLINTER	:09	XV Series SFX Library	URl

Figure 6.2 Search results for "big wood crash."

Although several of the effects listed in Fig. 6.2 look like something that will work, the only way to know is to listen, and this search engine only directs us to the proper place in our library. This is much slower as what looks like a great effect like XV 04-99 may not sound as good as we think it will. Some search engines allow for previewing right in the search engine.

One great way to search is if the audio has been encoded as BWF audio. Because BWF can contain metadata about the clip, this can be searched with any search engine. The problem is that for years, sound effects have been offered only as AIFF with no metadata. The search information is contained in a searchable database that directs the editor to the audio or in some cases previews the clip as a low-resolution MP3 or WAV.

NetMix Pro is a search engine designed to not only search BWF from metadata, but also to assist in converting AIFF libraries to BWF with the information from the database encoded as metadata. Prosound effects is offering all of the major sound libraries, over 250 000 effects, as turnkey NetMix Pro BWF media with metadata.

The Pro Tools DigiBase can search and preview any media including this BWF media. The comments column in the DigiBase will display up to 256 characters of metadata. AIFF and SD II can be searched as well, but because there is no comment metadata, searching by key words will only search the name of the clip. Often this works well, depending on the amount of information in the clip name.

DigiBase is used to search the entire workspace, in other words all mounted drives, for needed sounds. Keywords can be entered into the find area and any or all drives searched for possible sounds. As this takes time, the search is much faster if all sound effects and other sounds are stored in a dedicated area, preferably on a dedicated drive. This way the search can be limited to that drive and no time will be lost searching all drives and unrelated files.

Whenever a folder is opened in DigiBase, it is indexed. Indexing is the process of reading the media files, extracting the metadata and names for each file, then storing the data in a database so that it may be displayed in the columns of the DigiBase browser. Although this takes only a few seconds for a folder with 500 or 600 KB of media, it may take a minute or two to index an entire library if everything is in one folder. Usually folders are kept in their original layout from their original CDs. Although this means having hundreds of folders and subfolders, this is the better and faster system.

It is also possible to generate waveform plots for the sounds that are displayed in the workspace. The waveforms are created with the "create waveforms" command in the DigiBase toolbox. This can take some time depending on the amount of media in the folder. For this reason, you may want to index and calculate waveforms for your sound effects files prior to starting a project to speed the editing later on. If you select the topmost folder or even the entire drive and use the "create waveforms" command, all sound files in the folder or on the drive will be calculated. This may take all night, but once it's done it's done. The indexed waveforms will be instantly available whenever any folder is opened in the workspace.

Indexing rewrites the drive index and links this to the DigiBase. For this reason you should always keep your media in dedicated folders, preferably on dedicated drives. Never store media in the system folder or with the operating system or applications or any kind of application support. Organize the media into folders as it is placed on the drive. This is much simpler that organizing after the fact.

On HD systems and LE systems with DV Toolkit, a catalog can be created from indexed media. The catalog can be used as a "mini library" of media for individual projects. In the toolbox icon you will find "create catalog." Sounds can be searched in the workspace and dragged into the catalog. These sounds will not be imported and copied into the audio folder until they are used in the timeline. This way anything you feel you may want to use can be placed in the catalog, without spending time importing or copying unless you actually use the sound. There are several reasons to create a catalog on some projects. If several editors are working on different reels of the film, the supervising editor or sound designer can pick all of the sounds to use in the project. All editors will be using the "approved" sounds in the catalog. This helps assure consistency through the show. A catalog is also a great way to "tag" sounds as you find them without actually attempting to cut them in. This can speed the searching process. And on a TV series a catalog is a way to keep all of the sounds that are used in most shows in an easy to find place. Subcatalogs can be used for the individual shows. A few hours spent creating and maintaining such a catalog can save hundreds of hours of editing.

Database comments in a catalog are stored in the catalog database. Database comments, can be up to 256 characters, are searchable, cross-platform, and editable. To create a catalog of a folder and its comments, simply drag and drop the folder onto the catalog icon in the workspace browser. Pro Tools indexes the folder, then creates the catalog with the same name as the folder.

Catalogs are an excellent way to manage the sound effects media from the library. You can create a catalog for effects or several catalogs for types of effects and fill each with the sounds you intend to use in the edit. Then simply drag the clips from the catalog into the timeline. All of the files referenced by the catalog will be copied into the session audio folder.

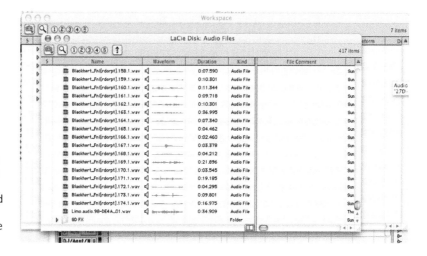

Figure 6.3 The Pro Tools workspace can be used to search all drives for sounds and preview right in Pro Tools. Found sounds can be dragged directly into the timeline or region bin.

The Workspace is opened in the window menu. From here you have access to all drives and all folders. After a volume is indexed, it is instantly available for searching or previewing. If the folder is closed before indexing is complete, the process will resume as soon as the folder is reopened; however, calculating waveforms goes on in the background even if the folders are closed. If the contents are altered, the indexing will be updated without reindexing all of the other media. If folders are kept small, the process goes quickly as the folder is first opened and allows for indexing as you browse and edit. If you are indexing your entire effects library, you may want to do it overnight.

Searching is performed with the search function (magnifying glass). Clips can be dragged into the catalog or directly into any track in the timeline. This copies the media into the session's audio files and if necessary, translates the media into the current format.

Creating a catalog for "Loves Devotion Forever"

With all of the indexed sound effects libraries mounted to one of the HD systems, the workspace was opened. From the browser menu, (tool box icon) a new catalog was created called "hard effects." A new folder opened by this name.

Now the search begins for effects using the volume browser. Clicking on the search icon (magnifying glass) opens a search window that is limited to the effects volume. A key word or words are entered into the find window. As these libraries are AIFF, there is no metadata other than the clip names. This is where having the comments information in the effects database or even better, as metadata in BWF, would be a huge time saver. As it is the search will be by volume and clip names and will return many effects that will not work with the scene. This searching can be to any volume, even a CD inserted into the drive. If the library was not already indexed, each directory and subdirectory would need to be indexed taking quite a bit of time depending on how much is in the directory.

As usable sounds are found they are dragged into the hard effects catalog. When the hard effects catalog is opened all of the effects are present and searchable for editing directly into the timeline. The catalog is still linked to the sound effects volume. If that drive is unmounted all of the effects go off line. When the clips are dragged into the timeline each clip is copied or converted from the effects volume into the session audio files folder on the session volume.

6.3 Recording sound effects

Very often there will not be any library effect that will work for the effect and the sound will need to be recorded or created. Even if the library effect is good, it may not be the exact sound needed. Or it may simply not be unique enough. Recording an effect is not much different from production recording, except there is no need to make concessions for the production crew and their needs. Much more time can be spent setting up the take and finding the perfect mic placement. One of the biggest challenges is finding a quiet place to work. Often the effects need to be recorded outside, and this means finding a quiet place with no birds, wind, airplanes, or traffic. Unless you want to travel to the Salt Flats to record (quite a drive but it works well), this usually means recording at night or in the evening before the crickets become a problem. This may be in an unexpected place such as an industrial park that is empty at night, or a cemetery. Once you find a place you will probably be recording there often to get to know the neighbors. Let them know what you are up to and also inform the police. They may want you to get a permit in which case you may want to find a different place.

Figure 6.4 Often the only way to get the proper sound is to record it in the field or studio.

Your recording may be very literal or it may be something you will be using to create something else. You may be smashing pumpkins with a sledge hammer, shooting arrows, chain sawing wood, recording automobile effects, throwing large rocks in a pond, or just about anything you can do outdoors. If the sound is of something that can be recorded on stage or in the studio, then that's the best place to record it. Many times you will be outside because you need to go to the sound. You may be after horses on a ranch, a tractor plowing a field, animal sounds, or whatever the scene calls for.

As always, think in terms of components. You may want to place a mic under the hood of the sports car and get just the engine sounds driving. Then record inside the car's interior. Then bys, ins, tires on dirt, tires on pavement, and so forth. Don't try to put everything into the recording at once.

This is also true of stereo backgrounds. Try not to record the restaurant. Record the people in the restaurant in one recording, and the dishes and movement in another. If there is music in the background, it's useless. While you may want music in the final scene, this is music and there are rights problems, mix problems, and the director will want something else and then the whole thing is out. Even a railroad yard should not have any noticeable horns. It's better if they are on their own track. This way you can place them where you want, as many as you want and as loud as you want.

Don't be afraid to back away with the mic or get very close if that's what you think will work. The conventional wisdom is record somewhat close; you can add reverb but you can't take it away. Although backing out is risky, the results may be worth it.

6.4 Creating sound effects

Creating sounds usually starts with recording a sound. If you are going to turn a donkey bray into a dragon roar (Lord of the Rings), you will need a certain donkey sound that may require recording lots of donkeys before you have something to work with.

The classic alteration is slowing the sound down and or pitch shifting down. Although this is simple, it can be very effective. Editing is critical. It's not likely the cow sound you recorded is perfect. You may need to remove some of the middle and shorten the tail. Slow it a bit and it's a demon. Speed it a bit and it's a zombie.

One of the challenges is digitally modifying the sound without it sounding digital or electronic. Unless you are trying to create a digital sound for a robot voice, then it's fine. But a digital sounding bear just isn't going to cut it. Often the old analog devices can be a big help. Slowing the effect down on the Nagra is a very natural effect even if the sound is slowed 200%. Digital pitch shifting sounds electronic if taken too far.

The best way to avoid a digital or electronic sound is to use digital effects very sparingly. Rely more on editing, mixing sounds together, rerecording in strange environments like underwater or inside the metal tank or anything else you can dream up. The good news is that anything might work. No one has ever heard the sound of hot kryptonite falling in cold water or the voice of a mummy. The last person to hear the sound of a charging saber tooth tiger didn't tell anyone about it. This means you can have fun and be creative. Think outside of the box or put the mic in the box and rerecord the sound.

Let's look at some classic sound altering effects that can be done in Pro Tools.

Rereversing

Figure 6.5 Reversing and even rereversing the sound can be used to create unique new sounds.

A sound can be reversed using the audio suite plug-in "reverse." The reversed sound can then be slightly altered and then rereversed with the reverse audio suite plug-in. This can be used in conjunction with many effects; however, the classic effect is to slightly reverb the reversed sound before rereversing.

Key gating

One sound can be used to control the volume of another by applying a gate to the sound you want to control keyed by the other sound. Let's say you want to make a demon voice with something of cat quality. If you

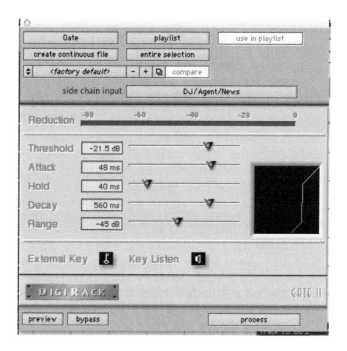

Figure 6.6 Using a gate with a side chain can alter the gated sound with the dynamics of the key sound.

spend some time recording and tweaking a cat growl-snarl this can be the basis of the voice. Place this in one track and apply a gate to that track. Place the recorded dialogue in another track. Hopefully the line already has a demon like performance, so that we are simply adding a subtle effect to help it out. Set the side chain input of the cat effect's gate to the dialogue track and select external key. Now the volume of the voice is controlling the volume of the cat effect. Blend the two together and you may have something that works in the scene. Or it may be just part of the effect; the modulated cat track can be further manipulated before mixing it back with the dialogue and/or the dialogue may need to be manipulated before it is used to gate the cat effect. This works well with rereversing. The dialogue can be reversed, used to gate the effect, and then the line and effect reversed so that the line is playing forward and the effect is playing backward. Now that's demonic. The SoundReplacer plug-in does something similar.

Speeding up, slowing down, and pitch shifting

Speeding up, slowing down, and pitch shifting are all related and can be used to dramatically alter a sound effect. Using the pitch shift AudioSuite plug-in with time correction only shifts the pitch. When used without time correction the pitch shift plug-in also speeds or slows the sound. Effects can sound electronic if taken too far, but slowing and speeding with pitch shift can be done on an analog recorder like the Nagra. Simply record at a high speed and play at a lower speed. When the speed is slowed dramatically, a high-pitched whistle can be heard. This is the recorder's bias shifting down into the audible range. This can be removed with a notch filter. Using an analog recorder, the sound can be shifted as far as you want without ever becoming electronic sounding.

Special plug-ins

There are many plug-ins designed to alter sounds into something unusual. Cosmonaut Voice is used to create an antique walkie-talkie radio noise. It can be used to "futz" other sounds to sound like they are playing through a radio or telephone. Lo-Fi also is used for this effect.

Some plug-ins create a fuller bass sound. With effects this is often done by slowing the effect down, while subharmonics can be created with Rectify and SansAmp. This can make thunder bigger and body falls sound like elephant falls. Waves Maxx Bass uses an old trick in a very new way. Instead of adding lower frequencies to the sound it adds higher frequencies that "beat" against the lower sounds. Smaller pipe organs have used this effect for over a 100 years. A 16-foot and 11-foot pipe played together can sound like a 32-foot pipe without the need of a 32 foot high room. Maxx Bass does basically the same thing. One huge advantage over subsonic systems is that the low frequencies are synthesized in the screening room, not played by the speakers. This means that this effect will even make small home screening systems sound bigger.

There are scores of plug-ins and thousands of ways to alter sound. One sound editor went so far as to turn his effects into MIDI samples and play the sound effects into the tracks with his guitar. It's OK to go a little nuts and creative as long as it works with the film and the producer and director like it. And like all sounds, leave as much control as you can to the final mix. The sound needs to be shaped in the mix to work with the scene.

6.5 Track layout

In editing the effects tracks use a logical track layout. Many editors like to put the more sync specific shorter effects, like a door slam, in the higher (low number) tracks, less specific sounds, like a car by or a tarp flapping in the wind, in the middle tracks, and backgrounds in the lowest tracks. (high number) This way when you have 50 effects tracks and there is a gunshot that is giving the mixer a problem, they can go right to the area in the tracks they know you would have placed it.

Backgrounds and ambiences

Every location should have a unique background sound (BG). This may be several tracks: wind, birds, stream, traffic, and so forth. The background helps to define the look and feel of the scene. Sometimes the backgrounds are "sub-mixed" in the final mix to make it simpler. The backgrounds are assigned to a group fader and the underlying BG tracks hidden. You have control of the BG through its fader and if the submix is wrong you still have access to, and control of, the individual tracks. Or the backgrounds can be premixed inside the session to a "nested" track and the original tracks made inactive. This is also called "ping ponging."

If the track count is becoming too high for a given system, many effects can be premixed in this manor. For example, the massive chandelier effect could be premixed into one effect. But always save the premix inactive tracks – in the final dub you may want to hear more of the plaster falling effect, and the only way to do that is to go back to the premix and remix.

Foley

Foley is a special kind of sound effect recorded on a Foley stage in sync with the picture and guide track. On large projects there is often one or even two Foley editors. More often, Foley is handled by the effects editor. Foley is cued early in the sound design, often as ADR and effects are being cued. The usual things recorded on the Foley stage are footsteps, cloth movement, and any sync effects to sync specific for wild recording or library effects. A glass of water is being poured and drunk, setting down a box of junk, a slap or punch, even something like a block of wood being sanded. Often Foley is used when the effect is hard to record outdoors because of background noise. The usual Foley crew consists of a Foley mixer and, because the chief product of the Foley is footsteps, two "steppers" with lots of shoes.

The Foley stage is similar to an ADR stage and often is used as an ADR stage at times, except a Foley stage has various floor surfaces and "pits," depressions in the floor that can be filled with sand, dirt, water, and so forth. There are also hundreds of props around a Foley stage that at times resembles a thrift store.

Recording Foley is a high gain situation: the sounds are often small and the mic placement is usually at some distance. Tiny unwanted sounds are a problem. Stomach growls, breathing, and unwanted movement sounds are often picked up by the mic and require retaking. A very clean mic pre is also necessary.

Figure 6.7 Foley walker Mike Crabtree walks the Warner Bros. TV show ER at Film Leaders studios in Burbank.

Figure 6.8 Foley mixer Tom Ruff recording Foley for *The Catbird Seat* at Film Leaders studios.

Some Foley mixers have their favorite plug-ins, mics, or other "tricks." Although Foley is mostly intended to be "naturalistic," it is always somewhat exaggerated, creating a subtlety surreal effect. And like ADR, it should never be right on mic or studio sounding.

Cuing Foley

Foley is cued in areas where there either are no footsteps or movement sounds, or where these sounds are thin and need to be enhanced. Often some shots in the scene were shot MOS (without sound) or they may have been recorded with lapel microphones that fall off dramatically and tend to only record the voice and not much else. Whatever caused the need, many scenes will require that footsteps and movement be rerecorded. Many films will be edited and mixed for foreign language replacement and may need Foley in all scenes.

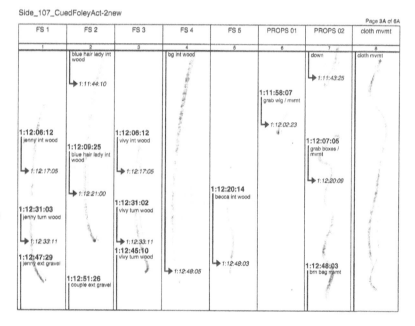

Figure 6.9 Foley is cued in the time honored way. This is also how ADR is cued and how the entire mix was cued when cue sheets were used in the mix. Each track shows the intended content, the incoming timecode, and the end timecode. This sheet was exported from ADR Studio. The purple lines were added by the Foley walker after each cue was recorded.

The ADR Studio plug-in from Gallery can be used to cue Foley as well as ADR. (For more information on ADR Studio, see Chapter 5). Specific sounds need to be cued individually. If the scene is a house falling apart in an earthquake, cueing "house falling apart sound" is meaningless. Moreover, there are things that just can't be done on a Foley stage. Although such a scene is a project for the effects editor, the Foley walkers can help with tightly synced details. Within the mayhem of this scene there may be a need for small sounds that the Foley team can record: a bottle tipping over and rolling, a lamp rattling at different rates, a pile of junk mail falling to the floor, or any small detail that is sync specific.

Large operations use cueing software that can be loaded into a controller on the stage, however, a simple list or spreadsheet works fine. Some editors like to cue the sound as tracks showing how they would like the sounds arranged into a defined number of tracks, however; there is something of a standard format that is usually followed.

Often eight tracks of Foley are recorded. Footsteps (FS) of the principals are recorded into track 1 and 2. Depending on the scene, more of the principal footsteps may be recorded into track 3. Tracks 3, 4, and 5 are usually footsteps of nonprincipals. Tracks 6 and 7 are usually props. A gun being cocked or handled, handcuffs, pouring a glass of water, or whatever. The last track is used for movement of clothes, the cloth track.

Figure 6.10 Foley walker Jennifer Wetzel creates the cloth pass. The cloth movement track is usually recorded sitting holding a cloth or heavy fabric. All of the cloth movement (MVT) for all the characters is recorded into the cloth track.

There may be any number of Foley tracks; while eight is common, some films may require substantially more. And just as with effects editing, break the sounds into components. If the cowboy in the black hat walks into the saloon, walk his boots on wood in on one track and the jangle of his spurs in on another. Finish it off with the movement of his leather chaps on another track and you are ready to serve him a shot of Red Eye Rye on the props track.

Foley is always cut into separate Foley only tracks, never cut into an effects track. Mixing Foley is quite different from mixing effects because it is a studio recording and all at the same level. Here too it may be premixed or ping ponged to fewer tracks; it may even be premixed to one Foley track.

Figure 6.11 The footsteps tracks are the heart of the Foley session.

It is also common to record "wild Foley," sounds that can be used in the effects edit but are sounds that can be recorded by the Foley crew on the Foley stage. Wild Foley sounds may be difficult to sync to picture, but are easy to record on the Foley stage without picture. For example, a cheese grater swinging on the end of a rope to be used as a dragon tail swish or a soda can being smashed to be slowed and used as a component in a car crash. The wild Foley effects can be cued to be recorded after all of the sync Foley is recorded, or the effects editor or creator may come to the Foley session or the Foley stage without the walkers and record their wild Foley and other effects.

Figure 6.12 I-glasses can be used as a video playback for field Foley recording.

A duplicate of the project on a laptop with an Mbox has the advantage of being a complete portable Pro Tools studio with two mic inputs, headphone monitoring, battery operation, and video reference. It is possible to record effects or even Foley with this system anywhere, assuming you have a quiet place to work. Inside a cave, on a ship, in a restroom, or whatever the scene calls for. The Foley walker can use the laptop screen, however, that is a difficult neck-bender. Video glasses can make the Foley walker's job much simpler. I-Glasses from IO display systems are available in video input as well as SVGA that can connect directly to the laptop and are battery operated.

6.6 MIDI Foley

It is possible to Foley on a MIDI keyboard. This can be done using a hardware MIDI sample player like the Oberheim DPX-1 or sample player built into some keyboards. More often, it is done with software sample players that can be linked to Pro Tools via the ReWire plug-in. Effects samples are available in several formats; be sure the sample player you are using supports your effects samples.

Once the effects are loaded into the sample player and being controlled by the MIDI keyboard, each key will play a different sample. Several very similar samples of footsteps can be loaded into several adjoining keys, the Foley footsteps walked with your fingers. The keys and effects should be velocity sensitive so that the volume and impact of the footstep is controlled by how hard the key is pressed. Other Foley effects, such as shuffles and scrapes, cloth, and pouring water are impossible using MIDI.

6.7 Backgrounds

Backgrounds are critical to the success of the scene. Each scene and every location needs to have a unique set of backgrounds that helps to establish the feel of the location. The filmmaker and the art director have worked to give the location a certain look and feel, and the background sound must support this look and feel. Even if a new scene is in the same basic location as the last scene, the new location or time needs to have a subtlety different background.

The backgrounds not only support the look of the set or location, but also aid in continuity. Even the best blocked and edited scene tends to have noticeable edits, and usually the goal is to make the edits so smooth that they disappear, or become "invisible" as noted film critic Andre Bazan wrote. By playing through the scene and across the edits, the backgrounds help define a continuous space, and a space that often extends beyond the camera's eye.

On stereo projects all backgrounds should be stereo tracks. Close-up effects can be centered or even panned left or right, but backgrounds need to fill the stereo field. Use a stereo mic or two matched mics and record to a stereo recorder. Many library backgrounds are stereo, however, many of the older ones are mono. Rather than avoiding all of these great backgrounds, they can be synthesized into stereo with several available plug-ins like the Waves S360 Surround Imager. Although this is an HD surround plug-in, it also offers mono-to-stereo as well as stereo to surround.

It is common for a scene to have several background tracks. Most scenes will have two or more backgrounds, and at the scene change the new backgrounds should come in on empty tracks so that the levels can be controlled in the mix later. If a scene has five tracks of birds, wind, leafs rustling, and whatever, and the new scene has five tracks of backgrounds as well, then you will need 10 stereo background tracks.

The key to backgrounds is not in the editing, but in the finding or recording of the perfect sound. There are a few things to keep in mind:

- Avoid sounds with foreground sounds like close car horns, a bell ringing, footsteps on the sidewalk, or anything that jumps forward in the background. Although these may sound great, they need to be controllable and should therefore be Foley or effects.

- Avoid backgrounds with walla or any kind of voices. These too should be in separate dialogue or effects tracks.

- Unless you want to get fired, sued and get no presents on your next three birthdays, avoid music!

6.8 5.1 Surround effects

The biggest difference in editing a surround project is that the backgrounds need to be four tracks, not just stereo. Rather than recording with a four-track recorder, this is most often done by recording two extremely similar stereo backgrounds using one in the front channels and one in the rear. Or the same background effect can be used in two stereo tracks slipped some 20 or 30s out of sync with each other.

Some sound design elements may be handled in the same way. As you work to create the sounds from some alien realm, you will want to take advantage of the 5.1 environment. Sounds can panned around in the mix, but four-channel spatial effects are just to cool to pass up. The same technique used for backgrounds can be used here too.

Some of the midground effects may get special 5.1 treatment as well. If the scene takes place in a landing craft on D-Day, the backgrounds of ocean and other landing craft should be four channels, as should the walla and engine sounds of the boat. But what about passing the aircraft? These can be mono or stereo and panned back to front in the mix. Or they can be four channel with the front stereo pair pitch shifted down so when the levels are matched in the mix, there is a Doppler effect (cool). We will be looking at mixer layout and mixing 5.1 in Chapter 8 "The Dub."

The foreground effects are generally kept in the front channels, usually the left and right but some mono effects may end up in the center channel. These are mixing issues, and don't impact the effects editing, other than to note that stereo effects will end up in the left and right; mono may be split between left and right or in the center channel.

The effects edit is creative and gives the editor and mixer a chance to really push the audio envelope to the outer edge. Like all aspects of sound it requires great recording, editing, and mixing. It is necessary for technical reasons, but more importantly it finishes what the rest of the filmmaking team started. It creates a world outside of reality, but inside the film.

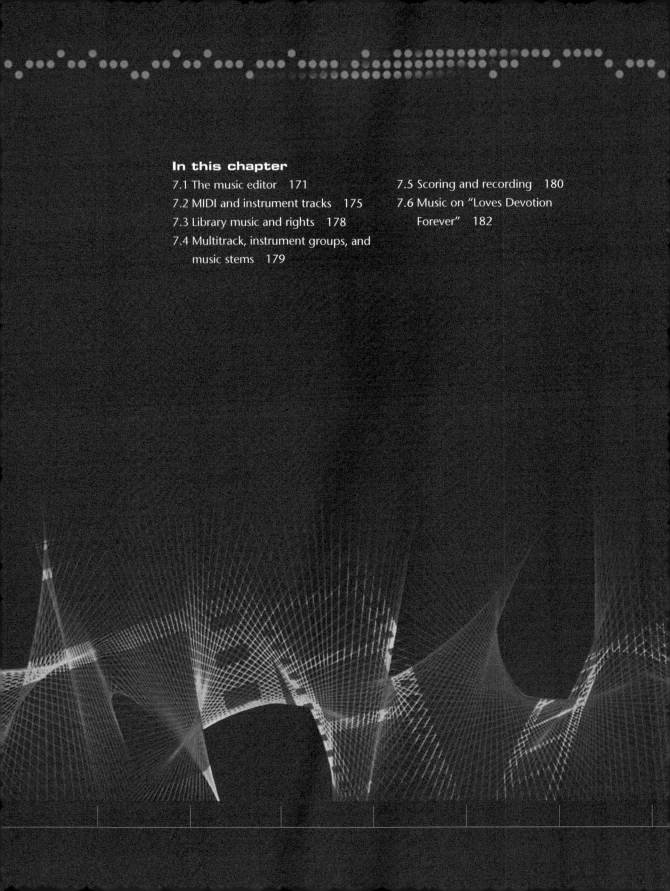

In this chapter

Music

Almost without exception, films have some kind of musical score. Even in the days of "silent" films, a score accompanied the film, played on a "Mighty Wurlitzer" pipe organ or even by an orchestra. Sometimes, a score was written; Charlie Chaplin even wrote his own scores for his films. Often, the score was simply a set of cue sheets; most organists improvised almost the entire film.

The techniques of music recording and editing are quite different from the techniques used in postproduction, but for the score, these worlds collide. The person most responsible for bridging this gap is the postproduction music editor.

Figure 7.1 Orchestra at the Utah premiere of *Birth of a Nation* at the Salt Lake Theater 1915. Photo courtesy of the Utah State Historical Society.

7.1 The music editor

Although the final score is composed to the locked cut, the music editing and composing often starts before the film is even shot. Although it is not necessary for the music editor to be a musician per se, it is important that the music editor understands basic musical theory and is able to read music. They also need to understand both Pro Tools worlds: video and film postproduction as well as studio music recording and production.

The music editor is often brought in at the very beginning; on some films, their work will be needed before musical scenes can even be shot. This may involve making click tracks for dance numbers or preparing playback tracks for musical scenes involving lip syncing or the simulated playing of musical instruments. This involves prepping DAT or other playback tapes for use on the set. Because playback on set is the responsibility of the production mixer, they determine the playback format of this "guide track." The director will also be involved in this planning as the playback tracks must be prepared based on the shot lists and camera blocking.

It is also common as part of preproduction for the director, composer, and music editor to start picking some of the temp music to be used in the picture edit. The composer will be writing some of the basic themes for the score in preproduction, and having some temp music already selected helps everyone communicate the "feel" of the scenes. Some directors, including Stanley Kubrick, even used temp music on the set to help the actors get the feel of the scene.

Click tracks are often used in place of actual music playback on set during production. This can be used for dancers who need to know the rhythm and speed; they can dance to these beats, which include a down beat, and the editors can cut to the click keeping everything lined up on the beats. Then, when the score is dropped in, everything should match up.

Click tracks can easily be made in Pro Tools. Create a new instrument track and apply the Click instrument plug-in. The tempo is set in the Transport window by shutting down the tempo ruler (conductor icon) and using the tempo slider. The click can be recorded into an audio track by setting the output of the click track out to an unused bus and creating a new audio track with its input set to the click track's bus. You can then record the length of click needed for the scene's playback. The music coordinator, the composer, and the director will determine the tempo of the click. The click can be used to set the tempo for dancers or musicians on set. The actual music will be composed later to this tempo and the picture editor will keep the beat in line across the picture edits.

In some cases, the actual music will be used as the playback track in production. The most obvious use of such a track is lip-syncing vocals. In this case, the final music is recorded first. Care must be taken to dub this music to the playback DAT or other playback format to ensure that the playback track is in sync to the final tracks used in the edit and the final dub.

When the final music is imported into a Pro Tools session, the section of music for each scene can be cut for individual shots in the timeline and recorded to DAT, exported or bounced for playback from digital media. If the production mixer is requesting DAT for the playback tapes, the dub to DAT should be made digital-to-digital (S/PDIF input) to ensure sample-accurate sync. Production playback can be from many formats, even directly from a laptop running Pro Tools. The call is left to the production mixer; their team will be responsible for playback on set.

If the project is to be shot on film at 24 fps (film speed) and finished or edited on video at 29.97 or 23.976 fps (video speed), there will be pull-up and pull-down issues. Films that are shot at 24 fps, edited at 29.97 fps, and then projected at 24 fps will be pulled down and then pulled back up in final finishing. Films that are shot at 24 fps and then finished on video at 29.97 will only be pulled down.

If click tracks or guide tracks are used in production, the guide and/or click track will be pulled out of sync when the production film and audio are pulled down. Even if the project is pulled back up for projection, this may mean that the music has been speed changed or resampled twice, which may have introduced undesirable artifacts.

Even in preproduction, it is critical to have the workflow designed so that the entire postproduction team knows when they need to pull-up or pull-down. If the final track will be encoded in Dolby, DTS, or SDDS, technicians from these companies should have input on the workflow from the outset.

If the workflow calls for the final dub to be done at video speed and then pulled up to film speed after the dub, and on projects that will simply be

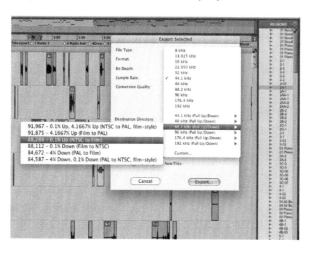

Figure 7.2 If the workflow calls for the final dub to be done at video speed and then pulled up to film speed after the dub, and on projects that will shot on film but finished on video, the guide or click tracks should be pulled up before being used on set.

finished on video, the guide or click tracks should be pulled up before being used on set. This way, when the film is pulled down in telecine to video, the film sync will match the original click or guide track.

Click and guide tracks can easily be pulled up from Pro Tools. Select all of the regions that you need to export pulled up in the region list and then select Export Regions as Files in the regions list pop-up menu. Select the proper pull-up sample rate for the onset playback (48048 or 96096 depending on the playback system) and export this media at the new faster speed. When pulled down later in telecine, this should match the original speed. This system is not sample-accurate and should not be used to pull-up a finished mix to match the original film speed, but it can hold visually accurate sync for even the longest takes in production.

As soon as the film goes into production, the editing of the picture also begins. Everyday new footage or "dailies" are delivered to the postproduction team, cataloged, dialogue synced if necessary, and strung into basic versions of the scene. In some scenes, the temp music will be added into this very rough edit. The temp music will need to be edited, shortened, lengthened, and changed for other music, all at the direction of the editor and director. At the same time, the themes for the final score are still being developed, changed, and at times, added into the edit.

Just as with Foley, stringers need to be added to the picture and this is normally part of the music editor's job. Hole punches are also used to mark key points. As with Foley, these are done nowadays digitally by layering digital files of "punches" and "grease pencil stringers" over the image in the editing software. The composer will communicate where they need these cues to mark key points in the score so that they can conduct the pacing to ensure the musical "hits" land on the proper points in the film.

The composer also needs timecode. The windows on the film transfer or digital video transfer show timecodes needed by the picture editors. They are the timecode from the original videotape or film transfer, the keycode numbers off of the film negative (if the project was shot on film), and the timecode from the production sound. None of this is needed by the music editor or composer; however, the timecode of the edited show is critical. This and any stringers (or punches) must be added to the video to be delivered to the music team using the picture editing software. Both Avid and Final Cut Pro have the ability to add this new timecode window and even mash out the picture edit timecode windows. This window can be set to read the timecode from the timeline or generate a new timecode. Standard practice calls for this timecode to align 01:00:00:00 (drop frame) to the first frame of the actual edit after the countdown leader; however, some editors prefer this timecode to be aligned to the Picture Start frame on the countdown leader.

As the film is fine cut, which is to say in a more-or-less final edit, the film is most often screened to test audiences to see if the story is engaging the audience and where it still may need to be tightened up. Temp mixes or "temp dubs" will need to be made for these screenings. Usually, this is with the temp music still in, but the film may also have some of the rough score in. The edit is usually changed several times based on the test screenings and requires major reworking of the temp music tracks. Finally, at some point, the film or film reel is "locked," which to say that there will be no more major changes to the edit. Often, however, the edit is changed after it is locked. Although the sound effects and dialogue editors can use Virtual Katy to reconform their tracks to the changed edit, there is no software package that can rewrite the score to match the new edit. The music department is dramatically impacted by picture changes after the lock. Some changes can be made in editing, but, more often, sections of the score need to be rewritten and rerecorded.

7.2 MIDI and instrument tracks

Along with audio tracks, the session may contain MIDI and instrument tracks. This is most often used to record and play music from a sampler, the sample being recorded into the sample player. Most often, the sample is a note from an instrument. The sample can be quite complex; many samples are recorded and edited together to form the beginning, middle, and end

Figure 7.3 While a MIDI track and an instrument track look quite similar in the Edit window, the channel strip reveals that there are no audio inputs, outputs, or inserts on a MIDI channel as there is no actual audio created in the MIDI track.

at several volumes. The sample is played from a MIDI keyboard, much like a piano keyboard. As the key is struck, the beginning of the sample is played at the proper volume for the strength of the keystroke, the middle is played until the key is lifted, and then the end is played. Only the keystroke is recorded as MIDI information into a Pro Tools MIDI track. A MIDI track is not sound but, rather, a set of instructions to send to a sample player. These tracks then need to be recorded as sound into a sound track before the mix is exported. However, as MIDI, they are extremely editable. You can change the attack, even change the sample played. A piano becomes a saxophone.

Because MIDI is not audio, MIDI tracks do not have any audio inputs or outputs. The input and output selectors of a MIDI track communicate only with the MIDI interface. The MIDI code is sent to the MIDI sample player that can be a separate device linked via the MIDI interface or the sample player can be a software running on the same computer as the Pro Tools session. Either way, the audio from the sample player needs to be routed back to the Pro Tools session on an audio track where the audio can be worked with and mixed into the final mix.

A Pro Tools instrument track differs from a MIDI track, in that it contains the sample player as a plug-in. This way the MIDI keyboard can send MIDI code to the instrument track, and the MIDI code can be recorded in the track, but the actual sound of the instrument plug-in is output from the track as audio. Instrument tracks therefore have audio inputs and outputs and can have audio processing plug-ins attached to the track.

Figure 7.4 Strike is a great drum machine, yet still a basic instrument plug-in from AIR. MIDI commands are recorded into the track from a keyboard, the MIDI editor, or the Score editor (version 8). Audio is routed from a mixer inside the plug-in from sampling microphones located at the drums, overhead, across the room, and even a talkback mic. All these mics are mixed to the stereo bus, the stereo Pro Tools audio track. The individual drums can be played as can elaborate "styles," performances that can be altered on the fly or via MIDI editor.

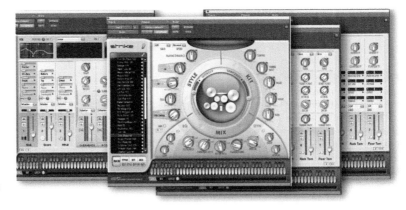

Figure 7.5 Several software packages are able to connect to Pro Tools using ReWire. In this case, Reason Adapted (adapted for use in Pro Tools) is open in the Pro Tools session. When Pro Tools rolls forward, Reason is interlocked and sends audio directly to the Pro Tools mixer. Reason offers a great sample player as well as a drum machine, full mixer, and several "old style" analog synth devices. It also has the coolest animated patch cables in the world!

There are several popular applications for use as plug-ins on instrument tracks or outboard as MIDI devices. Reason Adapted is often used with Pro Tools as it can be used directly inside Pro Tools on an instrument track. These applications are often called "sequencing" software because they can create a sequence of MIDI events. They are much more than this, however; they contain the hundreds of audio samples and control the entire "envelope" of the sound from key strike to key lift.

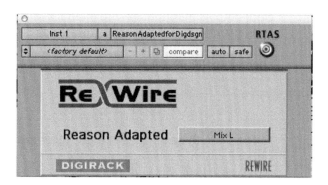

Figure 7.6 ReWire is used to connect third-party softwares to Pro Tools. Many applications can be linked via ReWire, including Reason Adapted, Ableton Live, MelodyneUno, and Sony's Acid.

177

There are many such applications that can link to Pro Tools. When used in an instrument track, these applications communicate with Pro Tools via the ReWire plug-in. Once installed and linked, these applications show up as plug-ins in an instrument track. When the instrument "plug-in" is attached to the track, the ReWire plug-in is attached and the separate application is launched, linked to Pro Tools through ReWire.

Creating these links is simple and automatic. Simply launch the ReWire application, then Pro Tools. ReWire will see the ReWire application and create a link. Shut down both softwares and the ReWire application should show up as an instrument plug-in in Pro Tools. When linked, all ReWire applications will need to be quit before quitting Pro Tools.

Figure 7.7 Not all these ReWire applications are sample players. Uno, for example, is more of an editing software. Uno converts single note music tracks into pseudo-MIDI; the audio is intact but is split out into something resembling MIDI commands. The individual notes can now be realigned to the beat and moved onto the proper pitch. Even the most off-pitch and off-tempo vocalist can be placed on pitch and on tempo.

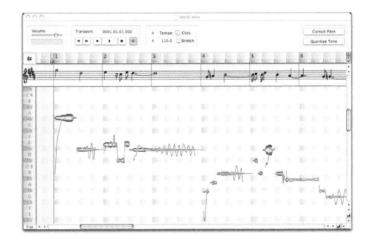

MIDI also has its own timecode format called MIDI Time Code (MTC). MTC can be used in place of video timecode to interlock MIDI-controlled devices like recorders and control surfaces. Many timecode locking devices have MIDI clock outputs making it easy to interlock MIDI hardware to your edit or sound design. Interfacing MIDI to hardware requires a MIDI interface device. This is often a simple USB device with MIDI in and out. Many Pro Tools interfaces also have built-in MIDI in and out.

7.3 Library music and rights

For decades, there have been music libraries with all types of stock music that can be used in a project. Historically, these have been collections of finished, mixed mono, or stereo music that can be used for a fee. These fall into two general groups, *buyout* and *rights available*. Buyout, also known as "free rights," is purchased outright. If you own the original CD or tape, you also own the rights to use it for any purpose other than selling rights to someone else. Rights available libraries contain music that can be used for a fee per use, also known as "needle drop" referring back to a time when these libraries were delivered on vinyl records. In this case, every use of every piece of music, even music extended through editing, must be licensed and paid for depending on the kind of project it is being used in. Whenever the project is re-issued or re-released, rights must be paid again based on the size of market and type of use.

In theory, all music is rights available. The original producer or person holding the rights can negotiate a fee to use the music. A library simply puts a fixed price on a collection of music so that there is little need to negotiate, and the price is typically much less.

This is often true of a recorded score as well. If the project is re-issued, a fee is usually paid to the composer for the music rights. If the music is re-edited for use in a different cut of the project, the same rules apply. It is normally part of the music editor's job to make a set of cue sheets after the final mix showing exactly what music was used where, if any extending through editing took place, and where the music originated from. These cue sheets will be used to calculate fees for final rights and if the project is re-issued.

Many newer buyout libraries are no longer collections of finished music but rather loops and samples that can be used in Pro Tools and other software to build complete pieces of music. This completely blurs the line between composer and editor.

7.4 Multitrack, instrument groups, and music stems

At times, the score or other music is delivered to the music editor as stereo or even mono tracks. Although this may be fine for a certain project, the best delivery is as multitrack recordings. This is not to say the unmixed tracks from the music sessions but, rather, music tracks that have been premixed into instrument groups as mono or stereo pairs. There may be many such pairs, strings, percussion, brass, vocal, leads, and so on. This way the feel of the music can be greatly altered in the final dub. Moreover, the placement in a 5.1 mix can be spread across the entire 5.1 spectrum rather than simply splitting the stereo between the stereo front and rear.

Delivery may be on tape or better, portable drive, but perhaps the best system for music delivery is via DigiDelivery. As new cues become available from the scoring session, they are uploaded to the server and an e-mail is automatically sent to the music editor with an encryption algorithm that functions as a secure key to download the cue to the music edit and final dub. This is very handy because usually the music is being recorded in a second location, sometimes thousands of miles from the edit or dub.

Ideally, the composer is present at the final dub because, with a multitrack score, there is so much control over the final feel of the music that their intent may be altered or even destroyed. The music editor acts as something of a representative for the composer and the score when the composer is not available. It is critical that the music editor and the composer be in creative sync and that the composer trusts the editor.

7.5 Scoring and recording

The score is recorded in a studio that is designed for music recording. The systems and studio designs used in postproduction are not appropriate for recording music. Yet many music recording studios may not have the proper equipment for video playback or syncing systems.

Figure 7.8 Fox's Stage one, the Newman Scoring Stage, named for the great Newman family, Lionel, Alfred, David, Emil, Maria, Randy, and Thomas, who together have been nominated for over 70 Oscars. A fixture at Fox, Alfred even composed Fox's signature fanfare. Photo courtesy of John Rotondi of Creative Audio Services who was part of the team that rebuilt the Newman Stage.

Not long ago, visual playback meant 35 mm interlocked projection. There were few facilities in the world that had such a system. Although some scores were recorded blind, without picture, most composers demanded film projection during the recording sessions.

The system included a room full of "dubbers," playback machines that could hold mono, three- or six-track audio. Usually, these were 35 mm, that is to

Figure 7.9 The interlocked projector was usually a massive theater system fitted with several interlock motors that allowed it to "rock and roll" with the recorders and playback machines. One of the great challenges with these systems was getting them to go forward and back at speeds greater than the usual 24 fps. At speeds greater than two times normal speed, most systems would shred whatever print was threaded on it.

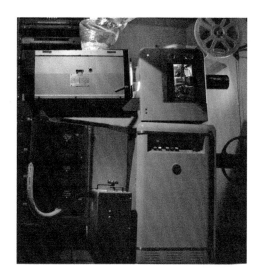

say, the magnetic film threaded on the dubber was 35 mm film base with magnetic oxide on one side. Other systems made included 16 mm, 70 mm, and even 8 mm systems.

Earlier systems used optical tracks on film. This system had the rather huge disadvantage that the optical recorder was actually a camera loaded with raw film. There was no going back and rerecording. If the recording failed, the camera was reloaded and another attempt was made on the cue.

Figure 7.10 Film dubbers (players) and recorders were interlocked with the projector and later even multitrack tape recorders. Although magnetic film has fallen into disuse, it is still used for some applications including film restoration. Pictured are two MagnaTec six-channel recorders and a double playback dubber.

These days video playback is so simple to set up that most home studios have it. It can be part of a portable Pro Tools recording system allowing just about any space to be a scoring stage.

Although many film scores are recorded on a large stage with a large orchestra, many are recorded on a basic system with only a few musicians because of the overwhelming cost of recording with a large orchestra. Most independent "no budget" films are scored with samples using MIDI software.

7.6 Music on "Loves Devotion Forever"

The temp music for "Loves Devotion Forever" was "composed" with Sonicfire Pro, one of the family of Smart Sound programs. While there is no ReWire version at this time for use within Pro Tools, Sonicfire is a stand-alone version that exports finished tracks in AIFF or BWF format. Depending on the input loops and samples, tracks can be exported as a stereo mix down or as separate tracks.

Loops were from the Richard Band library as well as the Smartsound library. Sonicfire Pro allows the editor to combine cues and tracks that can be combined seamlessly into music cues. Where most loop systems provide drum tracks, music beds, and bass line to which lead instruments can be added, Smartsound works with complete musical phrases that can be combined in different ways to "create" the music. Through multitrack recording and layering, the feel of the music can be greatly altered.

Also, one free rights cue was used from the T.H.E. Music Library, which the Brooks Institute owns for use in student projects. This cue is supplied in several mixes that can be combined and mixed in different ways to alter the feel and instrumentation of the cue.

The tracks were created and exported as separate files by instrument section. This created up to six separate stereo files on some scenes. Others were exported as stereo files only because the original loops were premixed to stereo. In some cases, the Sonicfire edits were preserved by editing in several tracks so that the individual tracks could still be exported as files. This gave improved control in the final mix.

Each cue was exported separately as AIFF with the Sonicfire session lined up at picture start. This exported a large amount of empty track at the head of every cue, but it held all tracks in sync and in proper alignment with each other. The tracks were then imported into the Pro Tools session with the "import audio to track" command in the file menu. This converted the files to BWF, and the new files were placed in a separate MX file within the audio files folder. The original AIFFs were also archived in this folder as well.

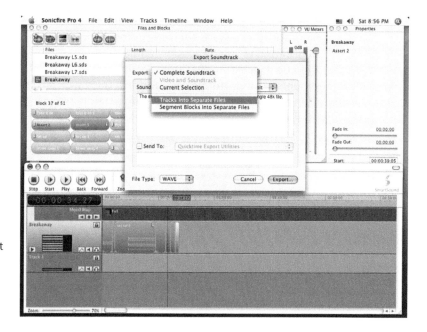

Figure 7.11 The export function in Sonicfire Pro allows for export of separate tracks as files when using multitrack sound loops.

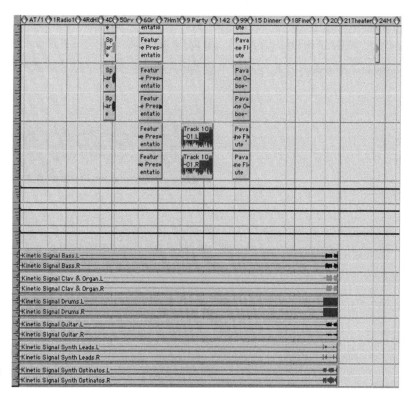

Figure 7.12 The import audio to new track item imports the clip and creates a new track placing the audio at picture start. All tracks for the cue are imported together – in this case, six stereo tracks.

The Sonicfire tracks had been placed in their own file within the session audio files, and these were then imported directly without the need to make any duplicate files in the session audio. The new tracks were created containing the cue as well as the long empty lead in use to place the tracks into the proper place in the project and in phase with each other. These tracks were then time-stamped using the "time stamp selected" command in the region list pop-up menu.

Figure 7.13 The six new regions are time-stamped. Note that the audio is only at the end of the cue, as the new audio has no matching timecode; the beginning is used only as spacer to keep the cue in the proper position. The tracks can now be cut down to the actual length and placed in the proper tracks.

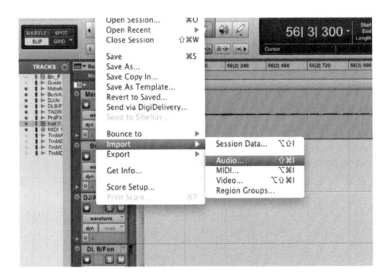

Figure 7.14 The time-stamped grouped regions can be placed in sync in the proper tracks in Spot mode.

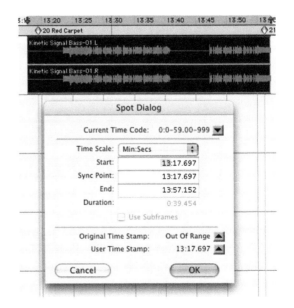

The time-stamped tracks were then grouped using the "group" command in the region menu. The lead-in spacer could now be cut away from the six stereo tracks and they could be moved to the actual music (MX) tracks. Because the audio was time-stamped, the spot function can be used to spot the block of 12 tracks into the proper place in the MX tracks by selecting the "user time stamp" in the Spot dialog window.

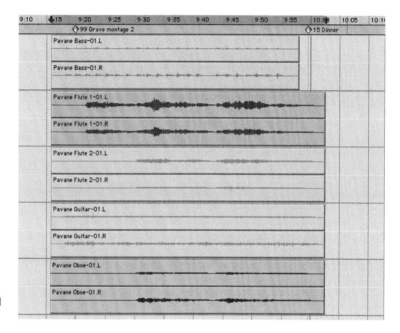

Figure 7.15 The final five track cue is ungrouped so that each channel can be controlled in the mix.

The tracks were then ungrouped to allow each track to be controlled and edited separately. The six tracks created by importing the audio to a new track were not deleted and the next cue was imported the same way.

Much of the new Sonicfire score was rejected in the temp dub and composer Mark Dunnett was hired to compose these areas. The score was recorded in Toronto, where this kind of work is somewhat less expensive without impacting the quality. In this case, the Sonicfire score made a great temp track, easily changeable and allowing experimentation in the mood and feel of the score. Even small changes in the score can dramatically affect the narrative. In this case, some of the Sonicfire score from Richard Band was put back in for the score of the film within the film. It was so exaggerated and over the top, it worked great as a score for Madeline's film. Such changes are so common as to be considered a normal part of the process.

The new score was also delivered as stems, giving extra control in the mix. All these stems were premixed using automation so that they would not need adjustment in the final mix unless the feel needed to be altered. Several such music remixes were necessary in the final dub to tweak the perfect feel.

Music editors and mixers live and work split between two worlds, music and postproduction, and they need to be well versed in both areas. They also need to understand rights and contractual obligations so as to not edit in or out something not covered in the contracts. Yet there are very few jobs more satisfying than bringing the score to life on a film, being on both the scoring stage and the dubbing stage. Although it is enormous work and involves long hours, some music editors even find time to have an outside life and even sleep, although normally the sleeping part is only done between film projects.

The Dub

8.1 Dubbing theory and techniques

Mixing the tracks, also known as *dubbing*, consists primarily of placing all of the sounds at the proper volume and equalization. Although this is the principal goal of the dub, the dub is also the last chance to fix any problems, place sounds in the stereo or surround perspective, and alter the sound to better fit the apparent environment or to fine tune the sound design. Mixing for postproduction is often more demanding than mixing music in a studio session. The music session often is only moved at points, and many channels tend to move together and movements are often small. In postproduction the levels, effects, and equalization tend to change all over the place all the time.

The sounds must work with the picture. This is not to say they must match the picture – in fact, usually they won't. Many sounds will be louder that the picture might suggest, others softer or missing all together.

It is critical that the sounds not compete with each other. Whatever sound is "featured" in the scene, the other sounds should take a back seat and let the featured sounds own the scene. This will often be the dialogue, but may be the effects, Foley, music, or backgrounds. Often the sounds pushed to the back will be so slight that they are barely audible, yet the mix feels stark without them. In many cases it will be necessary to have everything present, but in a very controlled proportion. In other cases only certain sounds will be heard. Although this is often the music being allowed to play forward with all other sounds muted, it may even be an effect, or even Foley, with everything else pulled.

It is important to manipulate the dynamics. At times the sounds will be huge and very dynamic. At times the dynamics should be compressed to support the dialogue or "fatten" sounds. It is the difference between loud and soft

Figure 8.1 Jim Corbett C.A.S., M.P.S.E. (left) and Charles Dayton C.A.S. on Todd AO's Glen Glenn stage one. Jim and Charles volunteer their time mixing the Kodak 35 mm Project films. Todd AO also sponsors the project with free stage time. Glen Glenn stage one is one of the largest and finest mixing stages in Hollywood.

that makes a sound seem loud or soft. If everything is loud, then the mix is bland; everything becomes annoying while nothing seems loud. In *Lord of the Rings* a trick was used to make very large sounds sound even louder. Just before the large sound everything else was pulled out for a fraction of a second. So rather than the large sound fighting with a large background, the large sound jumped out from silence. The fraction of a second of silence is not even perceivable, yet it makes the loud sound seem even louder. If you get in a car and keep your foot on the brake for next hour the car will not move. Machines are absolute and objective. But if you get on a horse and pull back on the reins for an hour you are likely to end up alone on the ground with a headache. Perception in living things is not analytical or mechanical. In this case all the horse perceived is that you were trying to hurt it. Perception is not absolute; we measure one thing compared to another. Unless you let up on the reins the horse has no way to know when you want to stop. And without soft there is no loud.

Feature films are mixed much more dynamically than television. For one thing the 24-bit theater systems can handle a much wider dynamic range than broadcast television. The question is often asked: what levels should I see on my meters while mixing? Assuming that the room is set up correctly for 85 dB at the mix chair (see Chapter 1 for more information on this), blood will be coming from your ears before you overmodulate. Therefore mixers mix to their ears, not their meters. The monitor speakers are normally set lower for television, between 80 dB and 85 dB causing the mixer(s) to mix at a somewhat higher level to get the same volume in the mix room. These higher levels can cause the mixer to overdrive the mix and so some attention to the meters is necessary.

Figure 8.2 Charles Dayton mixing effects for "Lost Hope and More" at Todd AO stage six.

The goal in mixing higher is to constrain the mix to a lower dynamic range without bringing up the noise floor. If the mix is constrained at a lower level it is closer to the noise floor resulting in a noisier recording.

The effects editor has selected backgrounds for each scene that support the feel of the scene. The mix, EQ, reverb, and compression must also support those environments. Just because the effects editor has cut in wind, birds, a distant stream and leaves rustling does not mean that all of these sounds are equally important and play at the same level. The environment on the scene should be supported by these sounds, and where some will be very forward, some will be hardly noticeable.

There are several ways to mix a film in Pro Tools, often divided into two basic systems referred to as "inside the box" and "outside the box." The outside the box system involves insert recording. This system is based on the oldest system used for mixing a film, where all the tracks are played together and the output of main mixing bus is recorded into a mix channel or channels. In its most basic form this system uses no automation, the levels are set and ridden on a control surface by the mixer(s) as the output of the mix bus is recorded onto one or more mix channels. Depending on the Pro Tools system and the complexity of the mix, this may require special hardware and software.

The first thing needed is a "PEC/direct" monitoring system. The monitoring needs to be assignable from the output of the mixing bus and the output of the mixed channels. Although these controls are present inside Pro Tools, it requires selecting monitoring using the mouse. A PEC/direct system needs to be easy to access and quick to use. Mouse clicks are too cumbersome for this task. The PEC/direct system allows the mixer to select what channels are being monitored and to insert the mix channel(s) into record.

The output of the main mix bus needs to be recorded onto one or more channels in the Pro Tools session. Although this can be done in the Pro Tools interface, it can be confusing and difficult. The simplest system is to use two Pro Tools systems, one for playback and one dedicated to recording. The recording system does not need to be extremely capable; even an LE or M-powered system will do provided it has timecode or MTC to lock it to the playback system.

The PEC/direct needs to have the following functions:

- Output of the playback system to the monitor speakers.
- Output of the recording system to the monitor speakers.
- Record for the record system.
- Transport controls for the interlocked systems.

Although most control surfaces have transport and record controls, none have a PEC/direct monitoring system. One can be constructed from common parts and can be laid on the control surface near the transport controls.

When insert mixing, the system rolls forward as the output bus is recorded to the final tracks. When a problem comes up with the mix, the system is stopped and returned to a place before the problem. The system is rolled forward as the mixer A-B compares the output of the main bus to the output of the mixed tracks. If any of the channels were not running from automation or recording to automation then the proper level(s) for those channels must be found before the recording system can be inserted into record. If the levels do not match, there will be a "bump" in the audio at the point of the insert. Using the PEC/direct system, the mixer quickly switches between the output of the bus and the playback of the mixed tracks while moving any sliders to find the matching levels. When no difference can be heard between the two, the record system is "punched" into record and mixing begins again from this point. All of this must be done quickly, usually in just a few seconds.

If the PEC/direct system is being used in conjunction with the transport controls on the control surface, then all that is needed is a small flat box that can sit on the control surface near the transport controls. On the box are two buttons, one that selects the output of the mix bus and one that selects the output of the recording system. The record system needs to be the master system so that the transport controls on the control surface control the entire interlocked system and the master record control in the transport controls can be used to punch into the record channels.

Now the mixer can shuttle to a point several seconds before they need to insert, press play, switch back and forth between the output of the playback systems main bus and the output of the recording system while matching levels on channel sliders with their other hand. When the levels match, punch the record button on the transport control.

Because even a simple mix can be too complicated to mix entirely in this way, some tracks are normally automated or premixed. Then the film is mixed in sections that are about 10 min long. Usually these are the "reels" of a feature film. (Features were historically edited in reels not to exceed 1000 feet. Although the actual reels of film no longer exist in postproduction, the concept of breaking the film into reel lengths is still used.) Usually the reel is rehearsed in a nonstop pass, as the mixer(s) try different levels and get the feel of the entire section. Then mixing starts from the beginning of the reel and continues live until a level is missed or any other problem comes up. Then the system is returned to a place several seconds before the problem. The system is then rolled forward while the mixer(s) find matching levels, and then the record channel(s) are inserted into record and mixing begins again at this point. When the mix is finished, the record tracks are backed up to separate files, and when needed, output to tape.

The advantage of mixing this way is that it's fast, and, depending on the skill of the mixer(s), very creative. The disadvantage of this is that if these mixing moves are performed live and not automated, there is no way to repeat the combination of moves. Changes after the mix become more difficult.

The other technique for mixing the film known as "inside the box" involves recording automation into every channel and then when the entire mix sounds right, bouncing the mix to disc, recording the mix directly to tape, or playing the mix back through encoding equipment or software to recordable media. The mix cannot be simply exported or backed up, as many of the audio processors, encodes and MIDI functions require real-time playback. When bouncing, the output of the main bus is recorded into one or more audio files that are the finished mix.

8.2 Premixing and premixing with automation

Often when a film project is very complicated it is necessary to premix some of the tracks to make the mix manageable. The goal in premixing is to reduce the number of tracks to be controlled in the final dub while keeping all of the tracks available to remix any premix if the levels used in the premix need to be altered to improve the final mix.

Automation can be used in place of predubbing if the system has the capability to play all tracks in real time. If the Pro Tools system is not capable of playing all tracks, predubbing will be necessary.

Predubbing can be performed in a variety of ways. One simple system is to set the output of all of the tracks to be premixed to an unused bus. Then create a new track in the same format as the mix (stereo, LCRS, 5.1, or whatever). Set the input to this new channel to the same bus as the output of the other channels. The selected channels can now be mixed and recorded to the new channel. The original tracks can then be made inactive, freeing up any resources being used by these tracks, and then hidden to free up space on the control surface and interface screen. If any problems come up in the dub with the premix, these channels can be made active and the premix channel altered.

If the Pro Tools system is able to play all tracks, and there is no need to free up resources being used by the premix tracks, and the goal is simply to simplify the mix, then the premix tracks can be automated, routed through a aux track to control the level of the premix, and all but the aux track hidden to free up space on the interface screen.

In this case, a new aux track is created, the input of the aux set to an open bus. Then the output of the premix channels can be routed to the same bus, and levels, plug-ins, and pans of the premix tracks automated. When this mix is sounding right, the premix tracks can be left active but hidden, leaving only the aux track to deal with in the final dub. If any adjustments are needed to the premix, the hidden tracks can be shown and touched up.

The most common channels to premix are backgrounds. There may be many of these tracks, and they can be premixed easily; it is easy to previsualize how these tracks will play in the final mix. One problem that often comes up is at scene changes. If the mixer wants the new background to come in at a certain level, then the outgoing scene will also be brought up to that level just before the cut. A solution to this problem is to premix the backgrounds to two channels, an A background track and a B background track making sure that the backgrounds do not butt up to each other at the scene changes. Now the mixer can set a level for the scene change without altering a large number of outgoing backgrounds.

Many channels are often premixed, usually effects tracks. It is exceedingly rare to premix dialogue tracks with the possible exception of elaborate walla effects. Foley tracks are often premixed as well, sometimes as one Foley track, or as two or three Foley tracks, for example, footsteps, movement, and props.

Sometimes the entire effects edit is premixed. In this case, however, this premix is usually part of the final mix; the effects are simply mixed first, then the mixer(s) return to the beginning and start adding in the dialogue and music. It is important when premixing to keep in mind how the entire mix will go together. There may be points where the sound seems thin or a transition feels unnatural, but keep in mind what will be going on in the other tracks. The transition may be covered by a huge music cue, or the thin sounds may be reinforced by other tracks not present in the premix.

8.3 The temp dub

A temp dub is usually performed. This is a good attempt to mix the tracks without any high attention to detail. The goal here being to check that all the sounds are here and the show is mixable.

Historically, the temp dub or temp dubs were necessary because the scores of tracks could only be heard on a mixing stage. The sound editors had never heard more than two or three tracks at a time until the temp dub. Only then could decisions be made as to what really worked and what did not. This is why often several temp dubs would be made.

With modern digital audio workstations it is possible to hear all of the tracks as they are being edited. Even if several people are working in several locations everyone can be working on the same project files via Digidelivery, or changes can be FTPed or even sneaker networked via portable drive.

This means that you know if a sound is going to work the way you want it to as you are editing it into the project. The need for a formal temp dub is diminished. Critical effects and music cues can be sent via Digidelivery to the director and producers for approval. With version 7.3 and higher, it is even possible to bounce short sections of temp dub with picture and send this for approval with Digidelivery.

On a film finish workflow, there are critical points along the way where a complete temp dub is needed. If a workprint is conformed this should be interlocked with the soundtrack and projected. This requires a temp dub to be transferred to DA88 or magnetic film. So in this case the temp dub may be nothing more than a touchup of levels and a transfer for screening.

It is also necessary to interlock and screen the first answer print on workflows where a silent answer print is made. And this, too, requires a temp dub.

On all projects there may be a need to screen a temp dub for producers, the director, network people, or whomever. In this case the temp dub may be done in a Pro Tools mix room by simply screening the Pro Tools session. More than likely this will be a more formal screening from videotape in a screening room with a good audio system. The temp dub will be bounced to disc, synced with video edit, and printed to tape. Pro Tools even offers a "bounce to QuickTime" for just this purpose. This is only an option when video playback is from QuickTime. In this case the mix and the QuickTime movie are combined into a single QuickTime movie. This is great for temping, but should never be used for delivery.

On many projects the temp dubs simply evolve into the final dub. On very large projects this is common as the sound editors and mixers may have access to a mixing stage through the entire postproduction. This was the case on the *Lord of the Rings* trilogy. The mix stage was dedicated to the project for the entire time the films were being edited. The online session was mixed for months as the sounds including the score were being recorded. At the end they gave the film a week of final adjustments and it was finished.

This is similar to how most small projects are mixed as well. As the no-budget guerilla filmmaker may be mixing on their own editing system, at some point they will give it one last set of tweaks and call it finished.

The "Hollywood" workflow is much closer to the older conventional workflow. In this case the sound may be mixed as it is edited, but at key points a temp dub is performed and screened until everyone in the decision loop is in agreement that the project is ready for the final dub.

Each sound can be approved or rejected in the temp dub process. Although rejecting sounds in the temp dub is common, it is expensive and time consuming to opt to reedit at the final dub.

8.4 The final dub

It all comes down to this. Now all the tracks are mixed as tightly as possible, every attention to every detail. Many of the decisions are subjective: how much background, how loud should the music be? Should we pan a sound as the character moves or not? There are no rules but there are guidelines. Never bury the dialogue. Keep the dialogue in the foreground over effects, music, everything. If you think the dialogue is forward enough, it's not. Go more. Don't get too big on effects and music unless it's a big film. Don't be afraid to go small with some effects. You don't need to always hear every footstep and shoe squeak.

Keep the intent of the filmmaker. Try not to let the mix turn into a committee decision. Figure out who is directing the mix – usually the director, but often this may be the producer – and let them make the call. Everyone at the mix needs to do everything to achieve the filmmaker's goal and the filmmaker needs to express his or her desires.

Figure 8.3 Juniper post's 5.1 mix room. Photo by Jeff Merritt.

Much of what is going on is technical. Are the levels going too high? Do we need to save headroom for later when we need to get much louder? There are standards for everything. The monitor speakers should be set at a level so that truly loud sounds are being heard loudly. The acoustics of the room are also critical. Mix at a reputable facility.

Most mixing facilities will want to start from scratch, but some may be interested in what levels have been used up to now and build on this. As always check with them as to their needs, but most likely they will want you to strip out all of your levels so that they can hear just what they have to work with.

Make a backup of the Pro Tools project so that if you need to get back to your temp dub you can. Then select the volume automation for each track, select all and press Delete. Remove any pan or mute automation the same way. Then remove all mixing filters.

Few mixing facilities use QuickTime for video playback. They want to keep all of the computer resources available to Pro Tools. Normally they use "machine control" where Pro Tools controls a videotape player and keeps it in sync with the Pro Tools session. They may also use virtual VTR or disc playback. If so, you will need to print your video to videotape in whatever format they use, or they will digitize this for their disc system. Also find out what timecode format they prefer. They will want 01:00:00:00 on the videotape to correspond to the beginning of the Pro Tools session, normally the first frame of the picture after the countdown leader. Or they may want it on the picture start frame of the countdown leader. You will probably need to have a professional videotape dub made to insure proper format and timecode placement. Even if they are using some form of digital playback, they will capture from this videotape to ensure that everything is 100% accurate.

If you are finishing on film, the absolute best workflow is to make the first answer print before the final dub. Then telecine this answer print to the proper videotape format with the proper timecode. This will ensure that the video reference is an exact match of any future prints. If there is a sync problem it could be because of an editing error, a conforming error, or any number of potential other problems. Rather than worrying about how it happened, simply fix it here knowing that this is absolutely the accurate picture.

If you are finishing on video, you can also mix after the online for the same reason. However, it is much less important on a video finish because if there is a sync problem it can be fixed by slipping the video shot rather than the audio. Because of this flexibility, in video finish the sound may be mixed before, during, or after the online edit.

8.5 Control surfaces and automation

Figure 8.4 The Digidesign Focusrite Control 24 control surface.

A control surface looks very much like a mixing board, and it is used exactly like a mixing board. But in fact it is more like a computer keyboard/mouse. It only controls the Pro Tools software and has very little built in function or active electronics. Some may have mic preamps or even a small monitoring mixer, but for the most part all of their function is provided by the Pro Tools system.

You may ask when one of the greatest advantages of the Pro Tools mixer is that it is virtual and therefore totally customizable, why tie yourself down to a huge hardware mixer? The advantage of such a controller is that many of the controls and sliders of the Pro Tools interface are now in the "real world" and can be quickly be touched and manipulated. Moreover, several controls can be manipulated at the same time and with better control.

The key to a good control surface is that it is very adaptable so that any custom Pro Tools mixer can be interfaced easily with the control surface. Many of the controls are obvious. A channel slider is always going to be a slider and every channel will have one. A pan pot is also usually present; however, it may be a single knob or slider, two knobs or sliders, or a 5.1 interface. If the control surface has a single knob pan, at times it will serve your needs, at times not.

Equalizers are a bigger problem: no two will have the same type or numbers of controls. Interfacing the control surface to an EQ plug-in may not even be possible on some control surfaces or plug-ins, but it certainly can still be controlled with the mouse. The better control surfaces have flexible systems to control EQ.

The transport controls are straightforward and easy to place on the control surface, as are automation controls, monitoring, shuttling and scrubbing, metering, and recording. A control surface is absolutely necessary for postproduction mixing; it can cut the time needed in half while increasing control and accuracy.

Figure 8.5 The Digidesign ICON control surface.

Automation with control surfaces is programmed more or less in the same way that it is with a virtual mixer. The controls on the control surface are touch sensitive; when in one of the automation write modes, any control that is moved will be written into the automation. Some controls are motorized so that the slider follows every move of the virtual mixer. Many knobs are not motorized but, rather, use LEDs to show the position of the knob. The display changes when the knob is moved or when the automation moves the control.

Figure 8.6 Digidesign ICON 16 channel fader strip module.

Some of the controls on some of the control surfaces are programmable and can be attached to many of the functions of the virtual mixer. The ICON D Control is available with customizable modules from which very elaborate, large control surfaces can be built. The ICON D Control is expandable to 80 fader strips per Pro Tools system.

Figure 8.7 Digidesign ICON producers desk – gear rack mount module.

It is not necessary to have all of the active channels present on the control surface. Control surfaces have "banking" controls that make it possible to move the faders across sets of channels in the mix. Only the channels actually being manipulated need to be present on the control surface.

Another advantage of a control surface, or several control surfaces controlling several Pro Tools systems, is that this allows the mixers to "insert mix," that is to say, mix with the control surface while recording the output as a final mix.

8.6 Outboard devices

As the needs of postproduction are often a bit out of the ordinary, it's understandable that some of the devices used in mixing for postproduction are also a bit "strange." Even though most audio processors are available as Pro Tools plug-ins, it seems that there should be no need for "outboard devices." Historically there were racks of devices near the board with processors such as equalizers, compressors, reverbs, noise gates, and even custom-made "futz boxes" and other strange devices preferred by the mixer.

Many of the digital plug-ins are attempts to recreate older outboard devices; they may even look like beat up old rack mount or home made devices on the computer screen (see Chapter 4 about plug-ins). And although they may sound very much like the historic device, they may not sound exactly like the device they are duplicating. Or the mixer may just love some strange device that looks like a reject from a government surplus sale. In any event, there are still times when you may want to connect an outboard device. And even though it may look like junk, it needs to be where the mixer can operate it, right out there in plain sight amid all of the high-tech 21-century computer-mixing tools. And it needs a patching system. These days some Pro Tools mixing systems may not even have a patch bay at the console, but if you want to use outboard devices, it is necessary to have a way to connect them to the Pro Tools system.

It is possible to use outboard devices just like a plug-in insert on the Pro Tools mixer. Inserts use corresponding inputs and outputs on the Pro Tools audio interface. For example, the device can be patched between output seven and input seven on the Pro Tools I/O box.

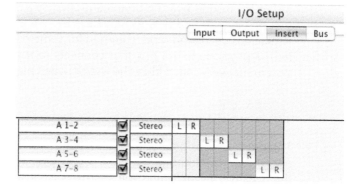

Figure 8.8 It is possible to set up the Insert I/O Setup to use outboard hardware devices as Pro Tools inserts in the Mix window.

The insert is now defined in the I/O Setup window in the Insert menu by selecting the IO channels being used with the hardware insert. The corresponding channels must also be disabled in both the input and output

I/O Setup. In Figure 8.9, the stereo in and outs are now accessible as an insert on the mix channels.

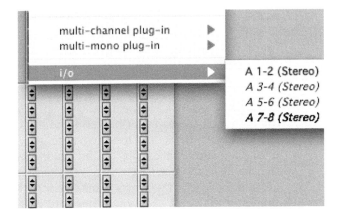

Figure 8.9 The hardware device is now available in the insert pull-down menu as an I/O insert.

Some hardware devices will connect to the Pro Tools system digitally. In this case the Pro Tools system must be the master clock for the digital device. Make sure to set your digital outboard device to external clock. See the Pro Tools manual for possible setups.

Some outboard hardware devices can be automated. They may connect by timecode, or they may use some other system, but it may be possible to connect them to the Pro Tools automation.

Just as the mixing board is often controlled by a mixing surface, some plug-ins that mimic outboard devices may have lost some level of control in becoming virtual. It is becoming more common for some plug-ins to now be controllable from a physical control box that connects to the control surface or Pro Tools system that very much blurs the line between plug-ins and outboard devices.

8.7 Levels and equalization

The most obvious goal of the dub is to set the levels and equalization of the edited sounds. Although the gross sound level needs to be altered in the mix, the dynamics of the sound will also need to be altered. A good recording will have wide dynamics, that is, a broad range between loud and soft sounds. Although this is one sign of a good recording, it may not be the best sound when mixed. Consider the sound of someone talking over a heavy noise floor, say a freeway sound. The dynamic voice recording has peaks and valleys, even within words. The peaks "punch through" the noise floor, but the valleys are lost under the freeway sound. This may make the dialogue hard to understand. If the dynamics are compressed to remove both the peaks and valleys, the compressed sound can be played up so that all of the sound punches through the noise floor. Reverb and other effects may need to be added to make the sounds fit the apparent environment of the scene.

Production audio

The production sound may be good, or it may be challenging with background noise, varying levels and varying amounts of room reverberation. This is often the hardest to mix, and it is normal for the most experienced mixer to be on the production tracks. Moreover, most of this sound will be dialogue, which is also the most listened sound and in many ways the most critical sound. Unless it is the intent of the filmmaker, the dialogue should always be easy to understand and the most featured sound. The dialogue will usually be equalized in an effort to give it more clarity and remove any low rumble and other undesirable frequencies. It often is also compressed to "fatten" the sound so that it can punch through the noise floor. This even helps when the noise floor is part of the production recording; even room reverberation is diminished to some extent when the sound is compressed.

Less is often more. There is a tendency for less experienced mixers to do too much: too many sounds competing for the listener's attention, and too much compression, reverb, and equalization muddying the sound. In terms of punching the dialogue through music and effects, more is often more. When the mix between the dialogue and everything else seems just right, many mixers will give the dialogue a tiny extra push.

Noise reduction is often necessary when mixing production dialogue. Unless the sound was unrecoverable, it would not have been replaced with automatic dialogue replacement (ADR). Sometimes some noise will need to be removed. If the problem is some unwanted frequency, for example hum, or if it is in the very low frequencies or high frequencies, it can be removed with filtering. A parametric or "notch" filter can do the job. If the noise is across the frequency range or at least across the voice range, it can be dramatically reduced with a broadband noise remover such as BNR.

A noise gate or the "strip silence" function may also be used. This lowers the audio track down or even turns it off if the dialogue drops below a set level. This can be extremely effective if there is a good effects background to fill the holes.

It is unlikely that you will want to place all of these plug-ins on every dialogue channel; some will need to be on the individual channels, but usually these are more manageable, use fewer computer resources and even do a better job if they are only on the output of the dialogue stem or on a send to the dialogue stem. In Chapter 3, we looked at how to set these up. Keep in mind that compression, a noise gate, broadband noise remover, or notch filter will not function properly on a send. The original uneffected sound will still be sent to the stem and the effects return with the noise reduction can't do anything about that. Noise reduction and compression must be inserted directly in the stem fader or on the individual production dialogue channels. Reverb and equalization intended to "sharpen up" the dialogue can be placed on a send, and many mixers feel they have improved control with it set up that way.

ADR

Although ADR is a dialogue, it is studio recorded and can be less demanding than production audio. The trick with mixing ADR is making sure it doesn't sound like ADR. The ADR mixer should have recorded the ADR so that it sounds more or less like the production, whereas in the dub it will need to be fine-tuned with level and equalization to sound exactly like the production dialogue. It will also be necessary to add reverb or "room" to some ADR. It may even be necessary to add compression to "fatten" the ADR; however, the ADR is most likely rather fat to start with.

Just as with production dialogue, these plug-ins can be placed on a send to the dialogue stem and route the ADR directly to the dialogue stem. It's unlikely that you will want to place ADR plug-ins directly on the dialogue stem, as this would add these effects to the production dialogue. You could also add an ADR fader and route all of the ADR through it and then to the dialogue stem. This would allow for the addition of noise reduction and compression; however, there should be no need to ever use noise reduction on ADR.

Walla

The walla may have been recorded or pulled from library sounds. It may need to be mixed into the effects stem, but only if it is totally nondescript and none of the words understood. More often it will be mixed into the dialogue stem. The walla needs to match the feel of the effects and other dialogue, and so is somewhat like mixing ADR. In most cases walla will be recorded at a more distant mic placement and may sound rather natural in its original state. More often it will need some EQ and reverb to get it to feel at home in the scene.

Sound effects

While the effects should consist of only high-quality sound they are at wildly wrong Levels because most of the effects were recorded wild without picture, or are from a library, and while they were recorded at a proper level for the recording, the level needs to be adjusted for this particular scene.

Often the effects will be very subdued, held way back under the dialogue or music. They may even be removed altogether and the music be given the entire track. At other times the sound effects will own the scene, even dialogue may take a back seat.

The effects are often compressed when they are very subtle and in the back, but when they are playing at full volume, they are as dynamic as the recording can support. The original recording may even be expanded in the mix. If a cannon is fired on screen, you may want to make the audience hear the full sound of a cannon going off in the theater. And the new 24-bit recording and playback systems can deliver it, too. Even if the effects that have been cut in are 16 bit, they can be dynamically expanded and mixed into a 24-bit mix. The effects will actually be converted to 24 bit when imported into a 24-bit mix, but while this adds headroom, it still requires taking advantage of this headroom by expanding the effects in the mix. They can be played louder, or they can be expanded with plug-ins like the Sonalksis Multiband Dynamics plug-in or even outboard devices like the DBX120A Subharmonic Bass Synthesizer.

The trickiest aspect of effects tracks is that there are often many of them. It may take 10 sounds in 10 tracks just to make the perfect sound for a peculiar effect. It is not uncommon to have more effects tracks than all of the other tracks combined. For this reason, it is often necessary to premix all of the effects before the final mix. This is certainly true on effects-heavy shows like large science fiction or fantasy films.

Foley

Foley is usually mixed more to the front than the picture would tend to suggest. The audio in film is almost always an exaggerated sound and there is often a sense that the audience has super hearing, able to make out every shoe squeak and paper set down. If the Foley is mixed too loud it will destroy the sense that this is actual production audio (*Kung Fu Movie*, anyone?). And while here too less is more, a bit more than natural is fine and most often the case. Often the Foley is not mixed by the effects mixer on multiple mixer dubs, but by the music mixer who has less to do and can handle the extra tracks.

Music

There is a misperception that the score requires little or no mixing at the final dub. Although it is possible to cut in the score as a simple stereo track, this is not the best practice. Delivering the score as a stereo mix is something like delivering the effects as a stereo mix. Yes, this works, but there is little or no control of the final mix and if the filmmaker wants something different, it can't be done. Although the music will need to be premixed before delivery, it is best if it is mixed to stems: strings, percussion, brass, solo, piano, or whatever stems will give some measure of control in the final dub. The goal of mixing from stems is not to alter the way the score has been written and performed, but to allow for subtle changes in the music mix to better serve the needs of the film.

Figure 8.10 The Waves S360 Surround Imager for Pro Tools HD.

It is also better to have the stems in a 5.1 mix as the instruments can be placed in different places in the theater. Stereo sound will need to be expanded to 5.1 with little control over what instruments end up where and even less control of them moving around at will. The Waves S360 Surround Imager is one such plug-in for imaging sound into a 5.1 mix with excellent results.

8.8 Mixing effects and futzing

Often the music is intended to be source music, coming from a radio or jukebox or whatever. This is often referred to as "diegetic" sound meaning that it is in the world of the narrative. In this case the music may be both futzed to sound like an old jukebox as well as equalized and reverberated to feel like the production audio. Sometimes music needs to sound like it is coming through a wall or floor. Rise the bass and smash the highs, add reverb and possibly compress for good measure, and it should sound just like a cheap hotel room above a bar.

Sometimes there is an intentional misdirection of what seems to be score that turns out to be source music. This is a bit tricky; if the mixer does anything too overt to the score it gives away the effect too early. In some cases the score is suddenly futzed into source music, but a gentle transition that sneaks up on the listener can be much more fun.

8.9 MIDI music

In some cases, some or all of the music may be on MIDI and not audio at all. In terms of mixing this can be more difficult, simply because there is no premixing of the instruments; each MIDI instrument is in its own MIDI track. True MIDI tracks will be brought into the mix on an audio track. The MIDI track will send its code to the MIDI sample player or other MIDI device and that device will return the MIDI as audio on an audio track. Although the audio track is treated the same as any other audio track, the MIDI device can be manipulated to also alter the sound. There is incredible control – the instruments can even be changed into different instruments. It is unlikely that the filmmaker will want to turn a piano into a saxophone at the final dub. However, if they want to, you can.

Some MIDI tracks may be Instrument tracks rather than pure MIDI. In this case the MIDI is handled internally in Pro Tools and the sample player is a plug-in. In terms of mixing this is really the same as working with MIDI tracks and outboard MIDI devices, but it is simpler to set up and does not require having the MIDI devices present at the dub. It does still require having the proper plug-ins installed on the mixing system.

Normally MIDI is premixed to stems as audio and then mixed into the final dub like any other score. But it is possible to mix with MIDI tracks directly in the final dub if access to the samples is desirable.

8.10 Surround mixing

Surround sound comes in many formats and flavors. One of the oldest and most often used is Dolby LCRS or left, center, right, surround. This format uses optical sound tracks that are encoded with Dolby analog noise reduction. Although this format is the most commonly used on prints, it is rarely used in actual projection. It was at one time the format of choice for 35 mm film, but it has been replaced by several digital Dolby formats, DTS and SDDS. Yet a LCRS track is still printed on the 35 mm print for theaters that do not support any of the digital formats and as a backup when the digital system is not working.

Some films are still mixed in LCRS and Pro Tools HD supports this format. Dolby even offers an LCRS plug-in for encoding and decoding LCRS, right in the Pro Tools session. Although any format can be mixed in Pro Tools, all of the other projection formats require encoding outside of Pro Tools. Most films are mixed in 5.1 surround, which can be encoded into a variety of formats and down-mixed to LCRS and stereo without any remixing.

Mixing using 5.1 can be quite time-consuming. Depending on the track layout in the sound edit, it may be necessary to further modify the Pro Tools mixer to accommodate the mix. In a worst case the edit may be stereo if the editors were stuck with working on LE or M-Powered systems that do not support 5.1. In this case the stereo edit will need to be changed into a 5.1 session. Let's take a look at that. First create a new session in 5.1. There are three 5.1 formats in Pro Tools:

- 5.1 SMPTE

- 5.1 Film

- 5.1 DTS

The only difference between these formats is in the speaker assignment. The format simply needs to match the I/O settings and the wiring of the monitor speakers. If you are sending audio to the right rear and it's coming out of the subwoofer, you're using the wrong format or wrong I/O settings. Find out how the system is wired.

Once you have a 5.1 session open, go to Import Session Data and import the stereo session. Import all tracks. This will alter the I/O settings back to stereo. Go to the I/O settings and set the I/O to 5.1 on the first six channels.

Now set the output of each channel to 5.1. The pan pots disappear and are replaced with the 5.1 pan window. Clicking on the small slider icon in the channel strip can expand this window. From here you can set the channel to the output channel or channels desired. If the edit was made with 5.1 in mind, all of the needed four channel sounds should be present in two stereo tracks each, or possibly four mono tracks.

As the editor(s) were working in stereo, they had no way of monitoring five channels, so they have edited sounds by visualizing the final mix monitoring both front and rear through stereo monitors. There's a bit of guesswork in this, sort of like painting in the dark. But it can be done and in fact is done all the time.

Another system for editing a 5.1 project is to work with a 5.1 template. It is possible to create a custom session template that keeps all the aspects of the session. This sets not only the format, 5.1, LCRS, stereo, or whatever, but also the mixer configuration, IO settings, track setups, and the preference settings. This saves you the time and grief of setting up the session every time you start a project. But, and this is very cool, even though LE (without the total production toolkit) and M-Powered systems do not support surround projects, because of the ability of Pro Tools to open sessions made on any Pro Tools system, 5.1 sessions can be opened on LE systems. Naturally there is only the stereo bus on these systems, so the 5.1 IO settings cannot be supported and they gray out routing everything to the stereo output. Moreover, the 5.1 session will crash the LE software if it is opened in 5.1; however, the 5.1 session can be imported into a new stereo session on the HD system using the "import session data" function in the file menu. But the session remains the same and will return to proper 5.1 monitoring when the session is imported back into a 5.1 session on an HD system.

It is not possible to create a surround session on an LE or M-Powered system, and you may not have access to an HD system to create the new surround session, but you can create a new 5.1 session on a stereo system by opening a 5.1 template.

The only real difference between a template and a session is that the template is laid out as a basic session with little or no media. The template can contain media however, so it can include a two pop, head tone, or whatever. But the idea is to create a layout for the session ready for editing or recording to the new session.

The template is also read only so that you don't accidentally save your new session over your template. After the session and template are opened, simply do a Save As and name the session. This allows the new session to record to drive and to be saved.

There are two ways to acquire a surround template: make one or find one. Often a mixing stage will email you one or you can go hunting for this choice Easter Egg on this book's Web site, several are included. If you find a template that serves your needs, but has many tracks you don't need, this is normal. A good template goes too far, containing all types of tracks, aux channels, and sends. You can always delete what you don't need, but you may not be able to create these tracks in the stereo session. Dolby offers several Pro Tools templates on their website for conversion of various formats.

Creating session templates

You can create a session template on a Macintosh simply by saving the session as a stationery pad. On Windows you create a read-only session. This session is now a template that you can open and save just like any other session.

To create a template on a Mac or Windows:

- Create a new session on an HD system in whatever format you want your template.

- Check and alter the IO settings and preferences as needed.

- Create all the tracks and the mixer needed in the template.

- Import any media or plug-ins you want in the template.

- Save and close the session.

- Locate the new session file.

- On Mac:

  - Go to Get Info by pressing Command-I.

  - Click on stationery pad.

- On Windows:

  - Right-click the Session file, choose Properties, then click in the Read-Only box.

This session can now be opened and saved as the new session on any Pro Tools system and doing a Save As in the File menu.

Any time you are working on an HD project on an LE system, Pro Tools will ask you if you want to convert the session to a two-channel LE session. When converted, all of the functions not supported in LE gray out including the IO settings. But the project can be edited in stereo, while visualizing what will be placed where in the surround mix later on the HD system. When the session is reopened on the HD system, again Pro Tools will ask if you want to convert the project; when converted, all of the grayed out functions come back online. There will still be a lot of tweaking and altering of the track settings and pans to get the tracks into the surround output, but the tracks are edited and ready to mix despite being edited on a stereo system.

8.11 Surround mixing theory

In the early days of stereo, music was often mixed with sounds in one speaker or the other, and sounds flew back and forth like a spastic chicken. Fortunately, the novelty wore off and engineers started working on creating a stereo prospective. When 5.1 came along, sounds were flying all around the room, dialogue was coming from the back of the theater, and the whole mess could be quite distracting. 5.1 is not Sensurround or Smellavision. It's not a gimmick; when used properly, it enhances the film and the story.

The key is to remember the film is on the screen at the front of the theater. This is a window to another world. Normally both the sound and image of this other world comes in through this window. Sound that seems to be coming from the back row in the theater is about as distracting as a cell phone ringing in the back row. It shatters the illusion that the viewer has been transported through this window into another world. The key is making the viewer feel like they are inside of this world, not in a theater. And if the cute little space alien's lines are coming from the back of a room full of people, it's hard not to notice that you are in a room full of people. 5.1 gives depth to the world on the screen and should not try to place the viewer in the middle of the scene.

For the most part the rear channels contain only music and backgrounds. This way the music is reinforced and given depth, and the environment becomes three-dimensional without pulling the viewer's attention away from the screen. The surrounds are mostly subliminal, adding depth without calling attention to themselves.

Although this may seem like the sides will need to be mixed to a lower volume than the rest of the mix, the sides can actually get very loud. It depends on what sounds are coming from the sides, and what sounds get to the listener first. This latter technique is based on the Haas effect. The Haas effect states that a listener will detect the direction of sound based on the first sound to reach their ears. If two cannons are fired at the exact same time, but a person hearing this is slightly closer to one, the sound will seem to come from that direction even if that cannon is the smaller of the two. Yet the sound will be very full, loud and have depth. The level of the delayed signal may be up to 10 db louder than the original at the listener and still sound like it is coming from the location of the first sound to arrive. The delay cannot exceed 40 ms or the sounds begin to sound like two distinct sound sources.

It's very difficult to say just when a sound will reach the listener's ear in a large theater, especially when the listener is not likely to be sitting in the center of the room. Part of the surround system requires careful timing of delays in the theater's sound system to keep the sound of the sides from reaching the listener before the sound from the front. This can expand the listening area in the theater to a very large area. This is not to say there are not crappy seats in the theater, but at least there is not just one good seat. For more information on the Haas effect, pull down a copy of "Über den Einfluss eines Einfachechos auf die Hörsamkeit von Sprache" by Helmut Haas. Or you can just take my word for it.

All of these effects can be used to create a very full surround mix without ever calling attention to the theater space. The Waves 360° Surround Tools offers several great tools to create a very spatial three-dimensional mix that never calls attention to itself. This can also be done just with levels and stereo plug-ins, just delaying the music in the sides by a frame (41.6 ms) can create a very spatial effect in most of the theater seats. And due to the Haas effect, the sound still feels like it is coming mostly from the screen.

Low frequency effects

The .1 in 5.1 stands for the LFE or low frequency effects channel. Although we all tend to call this a subwoofer, it's not a subwoofer in the home stereo sense. On home satellite systems there are five small speakers and a sub. In a satellite system the five front and surround speakers are not full range, lacking any woofer at all. For this reason the low frequencies from these channels plus the LFE are routed to the sub, the system's only woofer. This is handled in the decoder-preamp and is tuned to the surround system.

This is not the case in a theater. In a theater the speakers are full range, often with a good 15-inch woofers. Each channel is simply routed to the proper speaker or speakers and the LFE is sent to the LFE subwoofer. This is a true subwoofer meant to reproduce bone rattling bass effects. Only effects intended to really boom at the low end should be sent to the LFE channel.

If you are mixing on a satellite monitor system, rather than full range monitors, the management of how much bass is sent from the five front and surround channels needs to be handled by a plug-in or external device that properly tunes the sub to the speakers and the room. If you mix bass into the LFE because the bass is lacking in the five front and surround speakers, in playback, on either a full range theater system or a home satellite system, this will produce way too much bass. The Waves M360 Surround Manager is designed for just this purpose. Although this is an issue of studio setup (covered in Chapter 1), it is critical to understand this concept while mixing to avoid over cranking the bass.

Routing sound to the LFE can only be done by routing the sound to one or more of the other five channels and then route the extreme bass to the LFE. This is because the LFE can only reproduce frequencies at the very low end of the spectrum. If thunder is sent to the LFE, it will not sound anything like thunder. At best it will sound like a low rumble. The higher frequencies need to be played through a full range speaker or speakers to keep their characteristic sound.

This is why the 5.1 pan window has no direct assign for LFE, but rather a slider that adds the LFE to the five speaker pan system. When a stereo track is routed to the LFE, the extreme lows are mixed to mono and routed to the mono LFE.

Figure 8.11 Jim Corbett mixing dialogue on "Loves Devotion Forever."

8.12 Mixing "Loves Devotion Forever"

The final mix was done at Todd AO stage 6 by Jim Corbett and Charles Dayton. This is the old Glen Glenn complex in Hollywood; sacred ground for film audio people. Thousands of films and television shows have been mixed on this site over the last 60 years. The Glen Glenn studios became part of the Todd AO company in 1990.

A temp dub had been made at the Brooks Institute mix room prior to the final dub. The Pro Tools session was split into two separate sessions. One contained all of the effects, the other the dialogue and music. Four systems were linked together: three pro tools systems and a fourth Mac running Virtual VTR for picture playback (see Chapter 1 for more on video playback and Virtual VTR). The other three systems were Pro Tools systems, one for each session and a third used as an insert recorder.

As Charles is a certified member of GAS (Garage Audio Studios), he took the sessions home and did some checking and premixing in his excellent 5.1 HD garage. The mix came off without any notable problems. However, one line disappeared somehow and was brought to the mix on a keychain flash drive known as "The Sword of a Thousand Truths." But other than that, the mix was smooth sailing, thanks to the skill of Charles and Jim.

Figure 8.12 The Pro Tools equipment room at Todd AO stage 6.

Mixing is a powerful tool in filmmaking, and while a good mix is only possible with a good sound edit, the mix is where it all either comes together or it fails. The mix is the final step in making a film, and is a critical step. When you are done here you are truly done (even if the wrap party was 6 months ago).

Figure 8.13 Todd AO mixers Jim Corbett (left) and Charles Dayton (right).

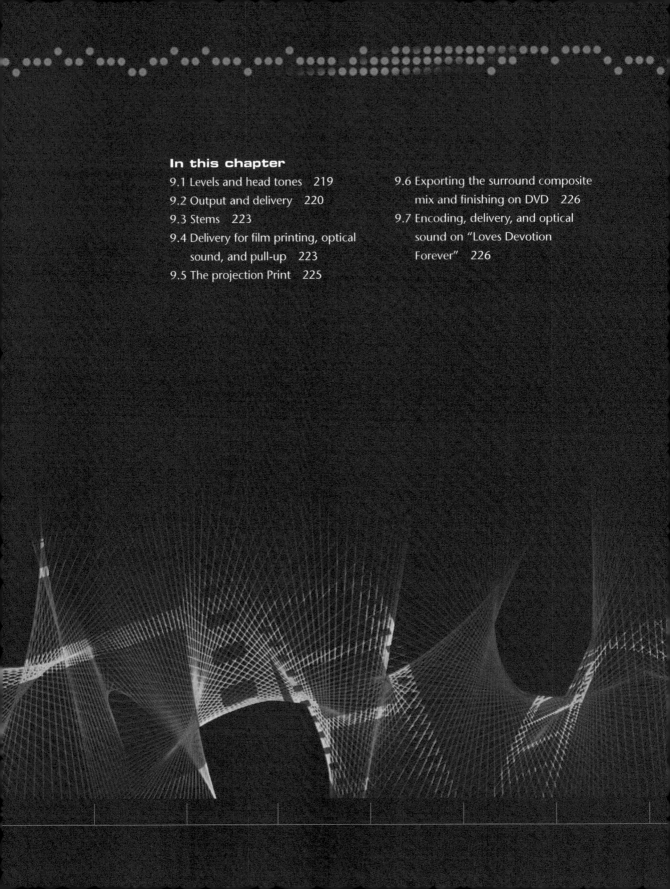

Output and delivery

9.1 Levels and head tones

Unless you are pouring cement, level is relative. When the police come to my house and tell me my music is upsetting the neighbors and is blaring at over 100 decibels I love to ask: SPL, dBu, or dBV? I then point out that each scale is based on a different reference, and 100 dBu would be right off my meters, so he must be wrong. He then points out that he can't hear a thing I'm saying because my stereo is blaring at over a 100 decibels and he is going to kick my . . . At this point, I know a civilized discussion of sound pressure level is pointless and attenuate accordingly.

When delivering the mix on tape, a reference tone is needed to keep everyone on the same page. If the digital audio is going to be copied bit-by-bit to another drive, no reference is needed. But if the mix is going to be dubbed to tape, from tape, shot to optical sound or any process where the level can be altered, the head tone will give everyone a way to check that they are transferring, encoding, or broadcasting at the proper level. Even if you are certain the project will be delivered digitally, it never hurts to lay a head tone, just in case some day in the future there is a need to lay the dub to tape.

Many tones and scales have been used, but for our digital delivery 1000 Hz at −20 dBFS is always used. Meters used in digital recording peak at 0 dBFS, meaning that they clip if the level exceeds 0 dB. And when digital recordings clip, the sound is horrid. Analog recording can exceed 0 dB and will clip at some unforeseen peak determined by the quality of the recorder, the tape, the bias, and a number of other factors. And when it does start to clip, at first it sounds great. Then it gradually becomes garbage.

Adding head tone to the session is simple. One system is to move the session start in the session setup window to 00:59:00:00 that adds 1 min to the beginning of the session. Be sure to use "maintain relative position." Record 30 s of blank media into one track. Now use the Signal Generator AudioSuite plug-in to convert this into head tone. The default setting on the Signal Generator AudioSuite plug-in is −20 dB at 1000 Hz.

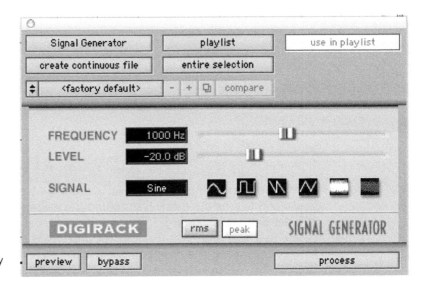

Figure 9.1 The Signal Generator AudioSuite plug-in can be used to lay head tone at −20 dB.

Another simple way to lay head tone is directly to tape. Don't extend the Pro Tools session; simply use the Signal Generator RTAS plug-in to send tone to tape and record 30 s before laying the mix or stems to tape. It's also a good idea to record a voice slate.

9.2 Output and delivery

Delivery of surround mixes can be from discrete WAV files for each channel for encoding in software such as Apple's Compressor, or the surround encoding may have been done in Pro Tools with Dolby or DTS plug-ins. If a matrixed format like Dolby LCRS or Pro Logic has been used, this can be recorded as stereo and a decoder can extract the surround mix from the stereo. Also, the stereo will play normally as stereo. Other encoded files will be delivered digitally.

Tape delivery may be on a variety of tape formats. One very popular and accurate system is to record the comp mix (and any stems) is on a digital tape recording system (DTRS) tape or compatible recorder with both the recorder and Pro Tools locked to house sync. Because of the two speed settings on the DTRS recorders, it can be used as both a video and film master, and with eight tracks it can handle even a 7.1 surround mix on a single tape. Delivery can also be on ADAT but this is falling into disuse.

Sometimes delivery is on videotape. This is more of a leftover of tape-to-tape editing, however, it is still used. Often this is Digi Beta, which supports four audio channels. It is exceedingly rare to lay a four channel mix, say LCRS Dolby, to this tape. Rather the four or even 5.1 matrixed encode can be recorded to two channels. This is called the left total (LT) and right total (RT), meaning that all of the left information is in the left and all of the right information is in the right. Often a stereo down-mix is also laid to the other two channels.

Digital delivery can be on removable drive, magneto optical drive, DVD ROM, FTP Internet delivery, DigiDelivery internet delivery, or digital data tape. Digital data tape like exabyte has been popular, but is falling into disuse. As these formats may join the dinosaurs soon, they are worth avoiding.

Delivery may include Dolby encode done at the mix. For 35 mm film projects, each film is granted a license from Dolby and must be encoded at the end of the mix by a representative from Dolby. The encoded mix can be recorded to DTRS tape and delivered to the optical sound facility. More often, the Dolby encodes are recorded to magneto optical drive. This is a rather old format, and has mostly vanished from the computer world. It was often used as a backup system. Magneto optical drive has been the standard Dolby delivery system for years. The drives are expensive, and the media is not on the shelf at Computer City. But the discs are robust, hold the data seemingly forever, and are not overly affected by magnetic fields.

Figure 9.2 Magneto optical discs come in several physical sizes and up to 30 Gig in storage.

Bouncing to disc versus layback

Bouncing to disc involves setting up automation for the entire mix and using the Bounce to Disc function in the file menu. There is also a bounce to QuickTime movie that also records the video if video playback is from QuickTime. This video is only used to create scratch exports to get approval from the director or producers.

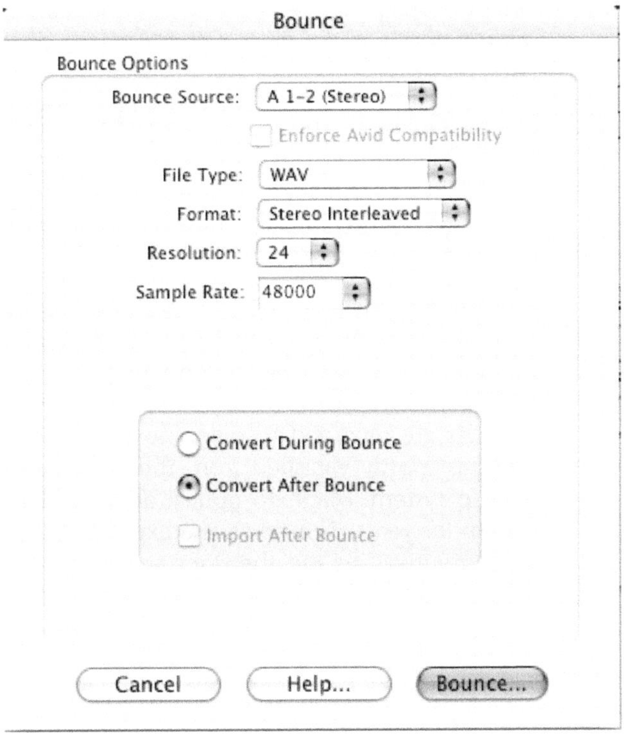

Figure 9.3 The bounce to disc dialogue.

Bounce to disc plays the entire project in real time and then creates the audio file. It's simple and the final output is one or more finished audio files.

Layback involves recording the session to tape, new tracks, or even another Pro Tools system that is interlocked to the session. It allows changing levels on the fly as the session is rolling forward. It is also possible to stop, back up, and remix sections. More details regarding this is discussed in Chapter 8 "The Dub." When laying to tape, the tape recorder should be interlocked to the Pro Tools system. There are several ways to do this depending on the tape recorder. The tape recorder can be the master and Pro Tools will follow it or Pro Tools can be the master with the tape recorder in chase mode following Pro Tools. The details regarding interlocks are covered in Chapter 1 "Pro Tools Systems."

Mixing to Pro Tools tracks involves creating a new stereo track, then setting its input to a pair of unused bus channels, then setting the output of the tracks to this bus. This track can now be record enabled and the mix laid back to this channel as levels are ridden.

If you find it necessary to stop, back up, and punch into record on these channels, they will not be complete self-contained audio files, but rather a series of inserted audio regions. They need to be made into self-contained audio files by selecting all in the layback track, and then using Consolidate in the edit menu. This will "merge" all the regions in the track into one audio file. These layback tracks can also be bounced to disc to make them into finished tracks.

It is also possible to mix to tape, in which case the finished audio is only on the master tape. This should be backed up as soon as the mix is finished. In fact, all deliverables should be backed up before the mix can be considered finished.

9.3 Stems

As the mix is going forward the mixing board is often set up to simultaneously lay the music, effects and dialogue stems off to another channel or channels. In some cases they will be laid off later. In some workflows they will be mixed first and then the "comp mix" (composite mix) mixed from the stems. When mixing from the stems, it is normal to make adjustments in the stems mix as it is going to the comp. If this comp mixing is made with automation, be sure to disable the automation before bouncing or laying back the stems so that they will not have these dips and gains. With 5.1 and other multitrack finishing formats there may be 5–6 music stems, 5–6 effects stems, and 3–5 dialogue stems. Stems for a stereo mix are usually one or two dialogue, two (stereo) music, and two (stereo) sound effects. Bouncing stems can be as simple as making all of the tracks not to be exported inactive, and then bouncing to disc. Or the stems may be routed to sets of busses and these bounced or laid back to a Pro Tools channel.

9.4 Delivery for film printing, optical sound, and pull-up

Before the "married" print (a print made with the sound track) can be printed, the audio from the mix needs to be shot to optical sound. If the project was edited and mixed at NTSC video frame rates, either 23.98 or 29.97 fps, the mix will need to be pulled up to match 24 fps. If the film was originally shot at 24 fps and was pulled down in transfer to video, then this pull-up is simply putting the sound back to its original speed. If the project was shot on high def digital at 29.97 or 23.98, the picture will be speed up to 24 fps when it is transferred to film and projected. In this case, the pull-up is syncing the audio to this new picture speed. If the project was shot on PAL video, the problem is more complex than simply pulling the speed from 25 fps down to 24 fps. This is a large enough speed shift to cause artifacts. The best workflow here would be to pull the 25 fps video down to 24 fps before recording, editing, and mixing the tracks.

Pulling up is more critical than pulling down. When we pulled the original sound down for syncing to 23.98 footage, we were slowing down by exactly 0.1%. Moreover, if we were off by some small fraction of this small percent, it rarely creates a problem as we only need to hold sync for the duration of the take, perhaps a few minutes on even a long take. The pull-up, however, is done to the entire film. We need to hold exact sync; even a drift of 0.0001% will cause the picture to drift out of sync. Although it is not best practice, it is not uncommon for pull-downs to be done with the equipment locked to several sync clocks. Pro Tools may be locked to its internal clock, a Nagra may use its crystal, and as these are both very accurate clocks we can hold sync through the take. But for truly accurate sync, every piece of the equipment must be locked to the same clock reference. Typically, a video sync generator is fed through a distribution amplifier and sent to every piece of equipment in the facility. This is referred to as "house sync." Moreover, though the pull-down was exactly 0.1%, the pull-up is a difficult 0.1001001%.

There are many ways to do this and the lab or sound facility will use their favorite system. Pro Tools can perform the pull-down and pull-up, and HD systems with SYNC HD can be locked to house sync. The Pro Tools session can be bounced to disc with pull-up set in the bounce dialogue window and the resulting audio file can be delivered on magneto optical disc.

Dolby LCRS is normally encoded on 35 mm prints as the optical track. LCRS provides four channels of high-quality audio. Encoding of any multitracks is usually performed before the optical transfer. Dolby encoding involves using phase angles to encode the various channels into the stereo optical track. Noise reduction and dynamic compression and expansion are also encoded, which dramatically improves the quality of the optical sound. A Dolby technician with encoding equipment is dispatched to the mixing facility after the mix is completed and the magneto optical disc or DA88 tape is encoded. The DA88 tape or magneto optical drive is then sent to the optical house to shoot the optical sound.

The optical track is a "picture" of the sound shot onto a narrow stripe just inside the sprocket holes of a black and white sound film. On 16 mm the optical is laid in the area just outside the standard 4 × 3 picture area extending all the way to the edge of the film. This area is used for picture in super 16 and sprocket holes on double perf film, so sound prints on 16mm are always single perf and never super 16. Super 16 must be finished digitally or blown up to 35mm.

Figure 9.4 The film strip on the right is a 35mm optical sound negative. When printed onto a positive print (the film strip on the left), only the white soundtrack stripes print through creating the two (stereo) undulating white stripes on the left side of the print. These can be played directly in stereo or mono. If Dolby encoding has been applied, they can also be played as left, center, right, surround (Dolby LCRS) sound in theaters. The center strip is the same print in 16mm mono.

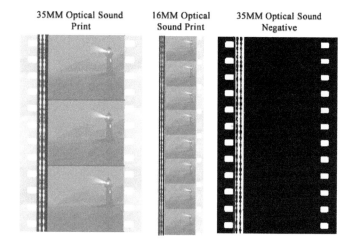

35MM Optical Sound Print   16MM Optical Sound Print   35MM Optical Sound Negative

There is no picture on the optical sound negative, it is only a black and white film negative containing the sound stripe image. It will be printed onto the answer print in one pass through the printer, the picture will be added in one or two more passes.

9.5 The projection print

There has always been a tremendous amount of unused space on the 35mm film print; the sprocket holes are well in from the edges and there is a fair amount of space between them. Also, the area used for the optical sound is large by "modern" standards. With scores of new digital sound formats now being used, every square millimeter of nonpicture film is being used to store audio information.

On Dolby Digital all six channels of audio information are recorded between the sprocket holes. New formats now record up to eight channels of sound. On DTS digital sound, the sound is recorded onto a compact optical disc (CD). A timecode is added to the optical sound in the form of a continuous barcode between the picture and the optical stereo track. This timecode pulls the digital sound into sync and constantly checks for sync. If there is a loss of sync or any problems detected by the system, it switches over to the optical sound until sync is restored. DTS is available for 16 mm; however, there are few 16 mm DTS theaters. On Sony SDDS, the soundtrack consists of an array of microscopic dots recorded onto both edges of the film just outside the sprocket holes. Like the tiny pits on a CD, these are the actual digital sound. There is still an optical sound track recorded in case of system problems and for theaters that cannot play SDDS. Dolby LCRS, Dolby Digital, DTS, and SDDS can all be encoded onto one print.

9.6 Exporting the surround composite mix and finishing on DVD

For DVD, 5.1 surround sounds can be encoded in Apple's Compressor as a part of the authoring process. For 5.1 Dolby the individual tracks are bounced as multimono and imported one at a time in Compressor. They are then compressed into a Dolby 5.1 AC3 audio file that is used in the DVD authoring.

As of late 2007, it is now also possible to encode DTS, Dolby Pro Logic, and Dolby Digital directly in Pro Tools using available plug-ins (see Chapter 5 about useful plug-ins). These are for DVD encoding only, not theatrical systems, which always require encoding by Dolby, Sony, or DTS.

9.7 Encoding, delivery, and optical sound on "Loves Devotion Forever"

After the mix was tweaked to a high state of finish, the Dolby technician arrived and connected his proprietary Dolby encode gear at the patch pay and the mix was encoded in 5.1 to a Magneto Optical disc.

The MO disc was hand delivered to NT audio, who exposed the optical sound film. This was delivered to Foto Kem Laboratories, who had already made three answer prints to check color. A release print was made and checked for problems. The film premiered at the Arc Light Cinemas in Hollywood 3 days later. Unfortunately all of our limos were in the shop so the cast and crew had to find their own way to the theater.

As the steps leading up to encoding and delivery are moving forward, at some point it becomes obvious that there is nothing left to do, that the project is finished. After months of around-the-clock pressure and making deadlines, champagne is often called for, but there is also a disquieting sense of decompression that can lead to a need for psychotherapy. Fear not! It passes as soon as you dive into the next project.

Figure 9.5 The Dolby encode equipment is patched into the Pro Tools system at the patch bay. Note the magneto optical disc on top of the case. The encode is recorded to the disc in real time through the Dolby equipment.

Figure 9.6 Dolby technician Bryan Pennington (right) works with Todd AO technicians Bill Ritter and Steve Hollenbeck to adjust the Dolby encode gear. The Dolby controls are on the desk in front of Bryan. There is also a Dolby meter bridge setup in the stage to monitor levels. The Dolby equipment is set to several tones played through the system.

Figure 9.7 The Dolby software is running from the number three system with the session still running on one and two. The encode must be done in one real time pass for each encode. In this case that requires two passes, one for Dolby Digital and one for Dolby LCRS analog. This pass also pulls the speed back to 24 fps for film projection.

Figure 9.8 Toasting the finish! Seated L to R, rerecording mixers Charles Dayton and Jim Corbett. Standing L to R, composer Mark Dunnett, executive producer Tracy Trotter, Dolby technician Bryan Pennington and yours truly, Dale Angell.

Appendix 1
Analog audio

Figure A1.1 The geniuses of the Edison Labs. The portrait shows Edison seated in center looking suspiciously like Napoleon. Fred Ott is seated on his right and Col. George Gouraud at his left. Standing (left to right) are the dapper W.K.L. Dickson, Charles Batchelor, A. Thedore Wangemann, John Ott, and Charles Brown. There is a phonograph and several recording cylinders on the table. Library of Congress photo.

Before we can understand analog audio recording, it is necessary to understand the nature of sound itself. Sound is simply vibrations in the air; sound as we perceive it is created in our ears as these waves cause small hair-like structures in the ear to vibrate at the same speed as the incoming airwaves. If these waves are vibrating too fast or too slow, human hearing cannot perceive them as there are no corresponding hair structures in the ear.

These vibrations are measured in cycles per second, also known as Hertz (Hz). Like a wave in water, sound waves have a peak where the level is higher and a trough where the level is lower. One cycle is a complete wave, for example, from peak to peak. The speed of these vibrations in no way changes the speed that the sound travels, which is more or less constant depending on barometric pressure. Because the sound travels at a fixed speed, slower waves are larger and faster waves are smaller. The speed of the wave is called frequency and the size is the wavelength.

The lowest frequency the average person can hear is about 20 cycles per second. This forms a wave just over 60 feet long. The highest frequency sound a person can hear varies widely from person to person. On average, this will be around 18000 cycles and a wavelength that is a small fraction of an inch.

While the size of the wave can be known, the actual "height" of the wave is relative. Is the tide in or out? This does not change the size of the wave; but if we are going to measure this wave, we need to keep our measurements inside of the wave, not relative to some fixed position. So we must state the reference if we are going to understand the measurement of the wave. If this were at a pier, we could say the peaks hit on the piling two feet below a line we painted there, or three feet above it. The line becomes a known fixed zero point. Or we may be on a ship so large that it floats at the exact sea level midpoint for that moment, and it too has a line on it where waves can be measured. This is a known relative zero point. Needless to say, you would expect to get very different measurements from these two scales even though they may be measuring the exact same wave.

Our ears work like the ship, otherwise sounds would be louder when it is warm and nice out and softer when it is storming. Ears register the changes in air pressure and not the actual barometric pressure. Our ears "float" like the big ship; and just like a big ship, they moan and pop if the "sea level" changes too quickly. If you don't believe this, try standing directly in the path of a tornado. No wait, that's probably a bad idea. Instead, try driving down a steep canyon and see if your ears pop.

The more powerful the wave is, the taller it is from peak to trough. More powerful waves are perceived as louder, softer sounds are smaller in peak to trough height. The height of the wave is called its amplitude and is measured in decibels (db). Bell Labs and Alexander Graham Bell created this measurement. A *bel* refers to a sound that is perceived to be twice as loud as some other sound. A *decibel* is one-tenth of a bel. Because a bel is referenced to a sound that the new sound is twice as loud as, it is necessary to also state what that reference sound is.

A common use of the decibel scale is sound pressure level (SPL). The reference sound on the SPL scale is the softest sound a person can hear. This is 0db SPL. Therefore, 10db SPL is twice as loud as the softest sound a person can hear, 20db SPL is twice as loud as 10db, 30db is twice as loud as 20db, and so on. The loudest sound a person can perceive is around 130db SPL. At this point, if the sound keeps getting louder, it is not perceived as getting louder, however, it hurts the ears more. This maximum SPL is known as the "threshold of pain." It is also known as a Jimmy Hendrix solo. (Yes, this dates me as an old hippie; Nine Inch Nails works, too.) The difference between the loudest sound and the softest is referred to as dynamic range.

Let's also look at phasing. In the real world, there is never a wave; there are infinite waves all interacting over time. When recording, we tend to look at the snapshot of the waves or the level at the microphone for that exact moment. However, waves move and interact, and depending on the position of that microphone, the net result will be altered.

Think of two waves moving together in the same direction, but separate. Imagine too that they are identical in size and shape. If the peaks and valleys line up with each other, the two waves are "in phase." If the peaks of one wave line up with the valleys of the other, the waves are 180° out of phase. If we had some way of laying the two waves on top of each other when they are out of phase, they cancel each other out and vanish. When they are in phase, they add together and make a bigger wave. While this new wave is "twice as big" if it were sound, it would be only 3 db SPL louder. This is because our ears are not linear but compress sound as it gets louder.

Microphones are very susceptible to phase problems because they have no brain to tell the microphone what to listen to and what to reject like our ears do. Our ear-brain is so good at this, we even use this to figure out the direction of sound. As the brain filters out phase angles between our ears, it can say, this sound should sound like this and it is coming from right over there.

If we use two microphones recording to two channels (stereo), this phasing information can be preserved, and when the listener stands between two stereo speakers, they can say "this sound is coming from right there." Even if "right there" is just a plant pot, it seems like that the sound is coming from there. But what if the recording is not accurate to the original phasing?

Try this: take a pair of speakers and play a single sound through both. It will sound like the playback is coming from a point directly in the center between the two speakers (if the system has more or less accurate phasing). Now, take the wires on one of the speakers and reverse the polarity. Just switch the wires + for −. Now, as the rewired speaker should be moving its cones out to make a wave, they move in making a trough. A listener standing between the speakers will hear weird interactions, and the sound may even seem to come from behind them at times. The brain just can't sort this mess out. The sound from the speakers is 180° out of phase.

But what if the sound is only a tiny amount out of phase, say 0.001 s? Low-frequency sounds are not altered that much, but very small waves are way out of phase, possibly even thousands of degrees out of phase. The net result is that the high sounds move around in playback, and usually sound like that they are coming from the speaker, which seems normal as this is where they are coming from. But if the phase was totally correct, even the cymbals should sound like they are coming, not from the speaker but from the original location between the microphones.

As this cannot really work in the real world, we cheat. 5.1 surround sound adds more speakers as if the sound seems like it is coming from the location of the speaker, this is more or less the goal. But if we could keep the phasing accurate, right down to the playback speakers, sound could sound like it is surround even if it is only stereo.

There have been several home theater and stereo systems that use this and "synthesize" surround sound from stereo (I have one made by Carver Sound, very fun). But this is not accurate, it just fakes in "placement." However, if you want to record and play sound so accurate in phasing that the listener can tell where the sound is coming from, even if it is over their head, you need to create a very accurate recording and playback system with almost perfect phasing and an analog of a human head. There are actually microphones shaped like a human head, some even have hair, that when recorded in accurate phase and played back through headphones (to remove phasing from the listener's walls), the result is amazing. Totally, direction-accurate recordings recorded on just two channels. The key to this is the head analog, a plastic head that shapes the phasing of the waves just as a human head does. But the sound itself can also be an analog of the original waves.

Analog or *analogue* comes from the Greek *analogos*, meaning something that bears an analogy to something else. In analog recording, some sort of model of the sound is recorded. This may be a magnetic image of the wave, an alternating electrical current, a physical model in plastic or metal, or even a photograph of the wave. The key to analog audio recording is that it uses a varying signal over time to create the exact same wave in some other recordable form. When recording a sound using an analog recorder, it is necessary to faithfully record a model of the frequencies and the dynamics. When the recording is faithful to the original sound, it is referred to as having high fidelity or hi-fi.

The upside of analog is that it can be very simple. The downside is that the recording is modeled in the recording medium, which can add in its own characteristics. Plastic records not only contain the model of the sound waves carved into the groves of the record but also the sound of any impurities in the plastic, which adds a characteristic plastic sound. Dirt and dust add their sound, as does the needle scraping along the record.

A brief history

Figure A1.2 Not exactly a Bluetooth ear bud, Bell's first telephone, 1876. Library of Congress photo.

The first use of analog sound was not a recording device but rather the telephone. There had been attempts at sending voice over telegraph lines, but none were successful. The telegraph is a digital device, sending only high and low pulses down the wire. Before the invention of the analog telephone, the hand key telegraph was already obsolete; automated systems were being developed to send high-speed code from keyboard devices that could be received and printed at the other end in real time. Digital fax machines had been sending images over telegraph lines since 1843. But digital audio was more elusive.

Bell's 1876 telephone differed in that it created an electrical analog of the sound in alternating current. The electrical waves were quite accurate models of the sound waves. The sound waves would strike a diaphragm, carbon grains pressed against the diaphragm would create an "undulating" current modeling the sound waves, and these were used to vibrate another diaphragm on the receiver that was connected to a coil of wire and permanent magnet. This diaphragm would vibrate the air recreating the original sound. This created a storm of new "analog" devices that were based on this undulating current.

Edison Labs invented the first analog recording system a year later in 1877. Edison himself came up with the idea while he was working on a digital high-speed telegraph code recording system that would use a rotating drum.

Figure A1.3 Original Edison Tin Foil Phonograph. Photo courtesy of U.S. Department of the Interior, National Park Service, Edison National Historic Site.

Edison realized that this digital recording device could be altered into an analog recording device.

The concept was simple: Bell's diaphragm could be used to scribe a groove into a rotating cylinder creating undulating waves in the groove. A playback stylus and diaphragm could then play this recording by tracking the undulating groove. Fred Ott at Edison Labs constructed the device in 30 h and it worked on the first test.

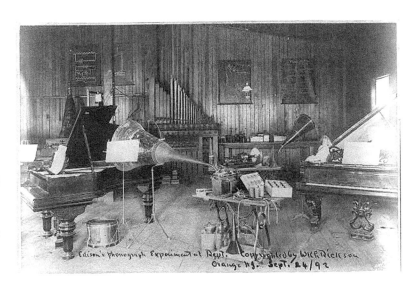

Figure A1.4 Edison's recording studio and test laboratory in 1892. Two pianos, a pipe organ, and several recorders can be seen. The doll on the keyboard of the piano on the right is a talking doll developed by Edison Labs and sold as a child's toy. Note the two boxes of blank recording cylinders on the table in the center. Photo by W.K.L Dickson, Library of Congress photo.

Years later in 1891, W.K.L Dickson of the Edison Labs constructed a motion picture camera. Edison saw little future in the device but realized that it could be connected to a phonograph for making sync sound films. While he still could not see any way to make a profitable business of this camera, he did feel that motion pictures of musicians would be novel and a great way to sell recordings. This was the birth of the "music video"! The problem for both Edison and Dickson was that they did not envision a projection system and there was no way to amplify the phonograph sound to fill a large hall.

The solution was the Kinetograph camera and the Kinetoscope viewer. These were silent viewers set up in "Kinetoscopic parlors" where patrons could view several short film clips for a small price. By the spring of 1895, Edison was offering Kinetophones, Kinetoscopes with phonographs inside their cabinets.

Figure A1.5 The Kinetoscope silent film viewer. As the device uses a "peep hole" in the top to view the film, only one person at a time could view the film.

True to Edison's prediction, the idea was a financial flop. During this same time, silent films were being projected in restaurants and clubs in Europe using the French camera-projector made by Lumiere and were a big hit. Edison soon started building projectors and continued to promote sound films while producing mostly silent films.

Figure A1.6 The Kinetophone was a Kinetoscope viewer with a phonograph inside. Note the tube earphones on the sharp-dressed patron.

Figure A1.7 This frame from an early Edison film shows W.K.L. Dickson (the inventor of the motion picture camera) playing a violin while two gentlemen dance to his fine tune. The large cone tube carried the sound to the phonograph recorder. Library of Congress photo.

Edison was forced to give up on sound films and never returned to that idea. In 1927, sound films came back, this time with amplified speakers and movie palace theaters. While these "new" sound films also used phonograph records for their sound playback, soon optical sound became the norm and is still in use today.

Figure A1.8 This sync sound projector by Edison was interlocked with electrical motors, but the sound was still not amplified and therefore this system was intended only for "home theaters." Library of Congress photo.

Figure A1.9 In later years, Thomas Edison turned to working with concrete and founded the Portland Cement Company. He patented designs for cast concrete houses and other unusual uses for cement. In this photograph, he is sitting with two of his phonographs, the one on the right is cast from cement.

Figure A1.10 The filmstrip on the right is an optical sound negative. When printed to a positive, the sound can be played on an optical sound reader. The strips in the center and on the right are film prints showing the positive optical sound and projection picture.

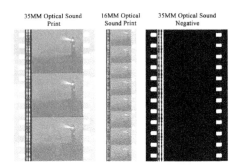

Optical sound uses the same photocell technology that made fax machines possible in 1843. The "electric eye" or photocell creates a small electrical current when struck by light. Sound was recorded on motion picture film using a system based on Bell's telephone. In fact, Bell was a big promoter of sound films. Bell had several companies: his manufacturing company was Western Electric, and his film dubbers and recorders sold under the name Westerex. An electromagnetic mechanical gate that opened and closed driven by the alternating current "model" of the sound was placed in the circuit from the carbon microphone. Light was passed through this gate and focused on moving unexposed film. When the film was processed, an image of the sound waves could be seen on the film. When light was passed through this film and focused on a photocell, a small current was created by the photocell. This was amplified and could be played through speakers, even in a large theater. This also meant that sound could be printed directly on the film print, making synchronizing and projecting no more difficult than silent films.

In production, an optical recorder was interlocked using electric selsyn motors to the camera. Both the picture and sound films were processed, printed, and could be put in sync by lining up the clapper sticks at the beginning of the shot and then editing with both sound film and picture interlocked on an editing machine.

Many tracks of sound, music, dialogue, and sound effects could all be edited one at a time onto scores of reels. Then all these tracks would be played interlocked with a projector and optical recorder. The tracks were mixed on a mixing console and the audio fed to the optical recorder. Because this system recorded directly to optical film, there was no chance to back up or redo the mix. Each roll of the film was mixed in real time, in one pass, and then processed and printed. Only then the mix could be played back to check for problems. Any redo requires beginning again at the start of the reel.

In the 1940s, the Germans were working secretly on magnetic sound recording. After World War II, the Allies "liberated" this technology. The concept is simple; the light gate in optical sound is replaced by a simple coil magnet. A moving iron wire is passed near the magnet and the fluctuating magnetic field magnetizes the wire as it passes. In playback, the wire is passed by the same coil, which now converts the magnetic fields in the wire back into electrical signal that can be amplified and played through speakers.

The wire recorders could not be interlocked to a camera or projector as the wire was notorious for stretching and slipping on the drive system causing it to loose sync. Soon a better system was developed. The wire was replaced with a plastic tape coated on one side with iron oxide (rust). The rust could be easily magnetized and the tape was much more stable. However, it still slipped on the drive mechanism and still expanded and contracted.

Iron oxide could also be coated onto film base, and the same basic interlock system used in optical sound could be used to record onto the "magnetic film." Magnetic heads (the coils of wire) were mounted on the optical dubbers with proper electronics to play or record from these heads, and the optical system could now also do magnetic. This meant that during the mix, the system could be stopped, backed up, played back, and a section rerecorded. It also sounded much better than optical sound. The final prints were still printed with optical sound, as they still are today. While there was a move to use magnetic sound on the release prints, the format never really caught on. Theaters were set up to project optical sound, but more importantly, the release prints had to be striped with oxide after the print was processed, and then the sound was recorded onto the print in real time. This slowed printing to a crawl and was expensive.

Recording directly to magnetic film was a hassle as the recorders were large and noisy and required line voltage to run their large motors. Small battery powered field recorders had been developed that used 1/4" wide magnetic tape, but again, as tape slips and stretches, it was unacceptable for sync recording in motion pictures. In 1960, a system was developed to modify a field tape recorder for recording in sync. The system worked like this: while the audio was recorded onto the tape, a second signal from the camera was also recorded. This was an alternating current that would oscillate at 60 Hz when the camera was running at 24 fps. If the camera sped up by 1%, then the signal sped up by the same 1% to 60.6 Hz. If the tape stretched, the recording of the 60 Hz signal was also stretched which caused it to playback slower. If 60 waves are recorded onto one foot of tape and the tape stretches to two feet long, it still has only 60 waves recorded on it. When played at the same one foot per second it was recorded at, the 60 waves are now spread over two feet causing, the speed of the waves coming off the recording to play back at 30 Hz.

The 1/4" tape was transferred to magnetic film using the same basic film recorder that before was used in production. In this case, the synchronous motor in the film recorder was driven by the 60 Hz signal recorded onto the 1/4" tape in production. As the sound was transferred from the 1/4" tape to magnetic film, the film recorder would slow down and sped up as the signal on the tape slowed and sped. If the tape was played back at the original speed, the film recorder would now match the speed of the camera, thereby holding sync. If the tape slipped or stretched, this would also cause the film recorder to slow or speed, still holding sync.

Another system was developed based on this one; in this case, a device called a resolver was used to control the speed of the playback tape recorder. When the signal on the tape slowed, the resolver would speed up the player until the signal was at the proper 60 Hz. This had the advantage that it never caught on fire. This is not to say that the earlier system was prone to catching on fire; but to drive the film recorder with the signal coming off the tape, this signal needed to be boosted to 110 V using a giant amplifier with tubes the size of mayonnaise jars. This generated more than sync, it generated lots of heat, and it could catch on fire if the right amount of dust and lint ended up in the amp.

The resolver made the transfer to magnetic film simpler and cooler. There is the old problem that the two devices need to be locked to the same clock if sync is to be 100% accurate (see Chapter 1). So if we want to hold sync for several minutes, the film recorder and the resolver need to be locked to the same reference. This is done by connecting the same 60 Hz line voltage to both the resolver and the film recorder so that they cannot shift their relative speeds.

This system is still used for analog recording in the film industry; however, the cable between the camera and the 1/4" recorder vanished years ago. Two accurate crystal clocks replaced it, one in the camera ensuring that the camera runs at 24 fps exactly and the other feeding an accurate 60 Hz signal to the tape recorder. Today digital recording is pushing analog right out the door. Digital recorders have many advantages over analog; however, some people still prefer the sound of analog.

Noise reduction and expanding dynamics in analog recordings

Two of the disadvantages of analog recordings are that the recording media add their own characteristics to the sound in the form of noise and the dynamic range is rather limited. It takes a certain amount of signal before the analog recorder can record anything. Very soft sounds are lost as they can't score the plastic of the record or magnetize the rust on the tape. Also, at the other end, very loud sounds can't be recorded, either. On a record, the groove becomes so wide that it crosses over into the next groove or the needle is moving around so much that it loses contact with the record. In magnetic recording, once the rust is magnetized to a certain point, it just can't be magnetized any more. So all analog recordings have a dynamic range from the softest sound to the loudest sound they can record and a meter to tell you when you are nearing the top of this range. This range is different for each medium. Digital recorders also have a dynamic range, but this is set by the bit depth of the recorder. In theory, it is possible to create a digital recorder with almost unlimited dynamic range, certainly greater than human hearing.

But most analog recordings need help. Humans can hear approximately 135 db of dynamic range. Optical recordings can only record approximately 20 db, cassette tape 30 db, a good phonograph 60 db. To boost these numbers, dynamic compression has become the norm. Most of these systems have been developed by Dolby labs and are used in virtually all theaters.

The compression scheme works something like this: while the sound is being recorded, the dynamics are compressed. As the sound gets louder, the system turns it down making it softer. It's like holding onto the volume control and turning the volume all the way up. Then when any sound is played, turn it quickly down. The louder the sound, the more you turn it down. The goal is to record a track with very little dynamic range, something that can be recorded even on optical sound.

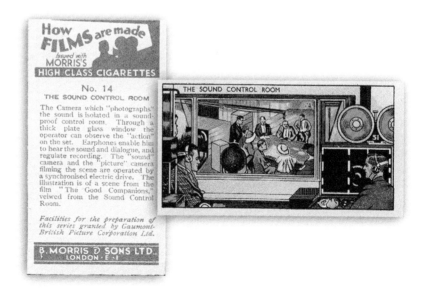

Figure A1.11 British cigarette trading card.

In playback, the exact opposite processing is applied. As the sound gets louder, the volume is turned up, as it gets softer it's turned down. If the recording compression is an exact mirror of the playback expansion, the original dynamics are restored, even though there is very little dynamic range in the actual recording. This can expand the dynamic range of a cassette or optical recording to approximately 70 db.

Dolby uses complex compression and expansion schemes that apply different amounts of processing at different frequencies. This not only captures a large dynamic range but it also removes much of the characteristic noise of the optical recording.

While analog sound has been for the most part replaced by digital audio, it lives on as a basic system for older theaters and a backup system in even the most modern theaters; and a few strange characters still hold onto their vinyl records and even play them, not just for the nostalgia but because they actually love the sound. They (we) are not likely to change their listening tastes any time soon. It is interesting to note that all the ultramodern analog systems (eight track, cassette, minicassette, VHS) have been the first to vanish, the surviving formats being phonograph records and optical sound, the oldest recording systems. These formats are 130 years old, and they can still give digital audio a run for the money.

Appendix 2
Digital audio

Note: If you haven't read Appendix 1 on analog audio, you should. This appendix refers to several topics covered in that appendix; anyway, there's some good information in Appendix 1 and it's fun to read and shows some wonderful uses for cement.

Digital audio differs from analog audio in that the sound is converted to a stream of data and then the data is recorded. The idea is a very old one; digital audio recording was written about and attempted in the early 20th century, but the digital systems of the day were mechanical and in no way could produce the data speeds to record audio. Digital audio became an idea waiting for technology to catch up to it. This finally happened in 1976 when Dr. Tom Stockham of the University of Utah made the first commercially successful 16-bit digital recording.

The concept works like this: analog electrical audio is sent to an analog to digital converter (A to D converter). The analog waveform is sampled at a high rate of speed; audio CDs use a 44100 sample per second rate. Each sample is a snapshot of the wave at that moment. This value is converted to a binary value; here again, audio CDs use a 16-bit binary word to represent this value.

As any good geek can tell you, 16 bits represents 65536 values. These values represent the total dynamic range of the recording. Because the analog wave has infinite variability, this requires "rounding off" the value to the next closest value. It can be seen in Fig. A2.1 that this creates a stair-step effect. If these steps are too far apart, they create a new waveform not found in the original. The best way to suppress these new waveforms is to keep the steps so small that the new waveform is above the high-frequency response of the recording and is therefore not recorded. This means that the total dynamic range of a 16-bit recording is 90 db. Not bad, but less than human hearing, which is about 135 db in dynamic range.

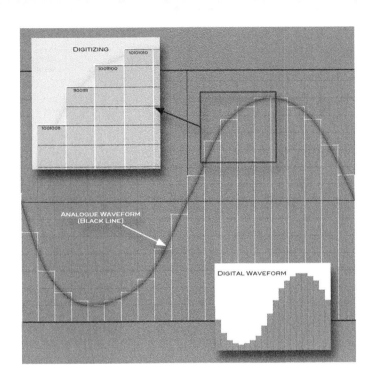

Figure. A2.1 In this illustration, the analog waveform (black curved line) is being digitized. The white vertical lines represent the sample clock, sampling the analog waveform 44 100 times per second on an audio CD. In the close-up in the upper left, it can be seen that each sampled value is being given a digital value. In the example, this is an 8-bit word, in audio CD, this is a 16-bit word. The digital waveform in the lower right shows a degree of "aliasing," the squaring off the edges of the samples.

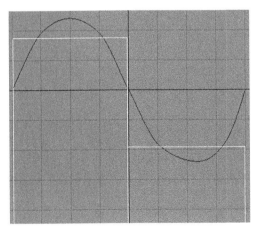

Figure A2.2 In this high-frequency analog waveform, it can be seen that it takes at least two digital samples to digitize a single wave. This is why the frequency response of a digital recording can never exceed half of the sample rate.

The frequency response is set by the sample rate of the recording. It can be seen in Fig. A2.2 that it takes a minimum of two samples to quantify one wave. This means that the maximum high-frequency response of a digital recording, in theory, is half of the sample rate. But, again, these waves are extremely complex, and the samples will square off much of these very high frequencies, and, worse, add in square waveforms that were not present in the original. This is called *aliasing* or *quantization error*, the squaring off detailed edges. These square edges are at a very high frequency, well above human hearing, yet create interactions that can be easily heard. To prevent this problem, high frequencies are filtered out of the original sound before the A to D converter converts it to a digital data stream. Moreover, the playback must be filtered at very high frequencies to remove these playback artifacts. This can be achieved with analog filters in the analog stream output from the digital to analog (D to A) converter that converts the data stream back to analog sound. Unfortunately, this adds phase shift in high frequencies that messes up the stereo imaging, which is to say the apparent placement of the sounds in a stereo field. Many sounds seem to move around from left to right because of this phase shift. A better way to "filter" these high frequencies is to oversample the playback. Although the recording was sampled at 44100, in playback, each of these samples can be read many times, perhaps 12 times. These samples can then be "anti-aliased," removing the square edges.

The filtering must remove all frequencies above half the sample rate; in this case, 22050 Hz. But the filtering can't simply cut off at 22050, it needs to ramp into the filtering; again on audio CDs, this filtering starts at 18000 Hz. Because this is the high end of human hearing, it was assumed that none of this filtering could be heard. However, this concept walks the edge of perception and recordings made at higher sample rates do sound better, mostly because of these aliasing problems.

The anti-aliasing uses a digital filter to remove high frequencies from the playback, and with the playback oversampled, this is now filtering a very high

Figure A2.3 In the waveform on the left, the original playback is over sampled; in this case, oversampling by 4. This takes the playback sample rate to 176400 Hz on an audio CD. This can produce an audio frequency response of 88200 Hz. When digital anti-aliasing is applied to this new sample rate, it tends to smooth the square edges of the original recording. Any analog filters used will now filter at higher frequencies that will not add phase shift to any audible sound.

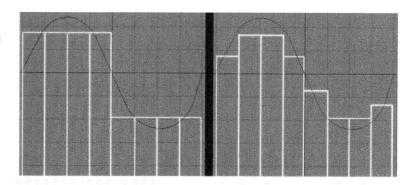

sample rate, as much as 12 times the original sample rate. Because of this, the filtering can function at one half of the new sample rate and therefore smooth some of the square edges.

Oddly, a small amount of noise is intentionally inserted into the analog audio before digitizing and sometimes more is added in post. It seems like the last thing that anyone would want to add to their sound is random noise, but it hides some of the digital artifacts. Not by covering them with noise but by giving the A to D converter something to work with at all times. If the same digital sample is recorded over and over, especially at the zero point, strange digital artifacts can be heard. The "dither" noise keeps something feeding the A to D converter so that it never tries to record a string of zero values.

Video has typically used a sample rate of 48000, which is an improvement over 44100. But in the last few years, 96000 has become rather common and is supported by Pro Tools LE when using the Mbox 2 Pro. Pro Tools HD now supports sample rates up to 192000 Hz. While the improvement to the high-frequency response is phenomenal, it is well above human hearing, so at first it seems pointless. However, it removes the need for oversampling in playback and filtering the recording.

Moreover, a bit depth of 24 bits has become common in professional recording. This expands the dynamic range to approximately 144 db. If you can hear it, 24 bit can record it. While none of this can be heard in the home on a 16-bit 44100 Hz CD or 48000 Hz DVD player, it is used in theaters using DTS, Dolby Digital, and SDDS. Even when these recordings are down converted to CD or DVD, they still don't have the same problems as a 16-bit 44100 recording.

Advantages

All digital recordings are based on a system that uses a clock and therefore can be easily used in film and video. Even on a CD player, the data coming off the CD is sent to a bucket memory and a controller. As the memory starts to fill, the controller begins sending data to the D to A converter for playback. A very accurate processor clock that is sending one sample set 44100 times per second controls the samples sent to the D to A converter. If the bucket memory starts to fill up, the CD can be slowed by the controller or the flow of data from the CD interrupted for a brief moment. If the clock starts to "over run" the memory, and the memory becomes low, the CD data can be sped up. So, while the data coming from the tape, CD, or digital file may speed and slow, the audio playback is extremely time-accurate.

While they are very accurate, no two clocks run at the exact same speed, and this will cause sync drift over time. However, many digital players and recorders can be controlled from the same clock. As we saw in Chapter 1, Pro Tools can be locked to any one of many different clocks. Many other digital devices can also be locked to the same external clock source. If a device is not controllable, it can be used as the clock source. When Pro Tools is locked to the Word Clock from a S/PDIF input, say from a DAT recorder, the clock controlling the speed of the 16-bit digital words in the DAT is also controlling the processor clock in the Pro Tools system. So, digital recordings are not only very sync accurate but also very interlockable.

Another advantage of digital audio is that it can be copied many times without degradation. Unless the copies are weaker or have other problems, the data in the recording is simply the data, a one is a one and a zero is a zero. Only if the data is damaged will the audio quality be reduced. Small amounts of damage will not be heard as the digital playback includes systems to rebuild data lost to dropout, but if the dropout is severe, the rebuilt data will not be exact and artifacts can be heard. If the damage is bad enough, the processor will mute the playback rather than playing random noise. However, if the copies are accurate, there is no generational loss in digital audio.

Error correction and compression

Dropout is corrected using parity check and other tricks to rebuild the missing data. Because the digital data can be shuffled and mixed with other data, several systems can be used to check for damaged data (dropout) and repair or interpolate a repair. Checksum error correction uses a fixed interval in the sample information. All the values of the samples inside the interval are added together. This number is recorded at the end of the interval as the first checksum. Then all the sample values are multiplied by their position in the interval; the second sample value is multiplied by 2, the third by 3, and so on.

Let's look at a simplified example. Let's say we have four samples in our interval. The values of the samples are 2, 4, 7, and 5. So, the first checksum is 18. Now the values are multiplied by their position. They become 2, 8, 21, and 20. The second checksum is therefore 51. Now let's say in playback we get values of 2, 6, 7, and 5. This checksum is 20, so we know something is over by 2. But which value is over? The second checksum is 55 or 4 over; 4 is twice 2, so we know that the second value is the errant sample. Therefore, it is reduced by 2.

Another system uses interleaving and interpolation. In interleaving, the samples are not recorded in order but are recorded in a predictable pattern. The first sample may be the 50 th recorded, the second sample may be the third recorded, the third may be the 32 nd value recorded. Now, if there is damage on the tape or digital file and 12 values in a row are dropped out, the missing information is scattered all over the waveform. If several values in the check-sum interval have been lost, there is no hope of the checksum error correction fixing the problem. However, because the missing samples are scattered all over the waveform, the samples on either side of the missing sample can be averaged and this value is used in place of the missing sample.

This type of recording is called pulse-code modulation (PCM). The pulse refers to the clock pulses, the code to the digital data, and the modulation to the waveform. Each pulse is recorded along with the error correction information. These recordings are more or less uncompressed, every sample is recorded as a 16- or 24-bit word.

But some digital audio is compressed to fit more on a storage system or to reduce the bandwidth of a transmission or a recording. Compression is used to reduce the size of the data, some information can be restructured in such a way to make the file smaller, and then restore the information in playback. Other information can be encoded inside the compression scheme. For example in Dolby AC-3, up to eight discrete channels of audio can be interleaved along with information about which speaker in a surround system will receive which audio stream.

Some compression is lossless; all the audio information is reconstructed in playback. Most systems are not able to restore all the information and are therefore "lossy." Because digital audio is a stream of extremely varied wave-forms, many of the compression systems used in other data storage formats will not work well as even a small amount of compression can be heard. Many audio engineers simply accept this degradation. The assumption is that most MP3s, for example, are only played on portable players with cheap head-phones and can therefore be compressed in spite of quality loss. With MP3, the user can download different files with more or less compression depend-ing on their needs and what they find acceptable.

There are a couple of compression schemes that reduce bandwidth and file size without creating noticeable artifacts. One system is rather similar to the oversampling scheme used in CD playback. While frequencies that were never recorded can never be played, a wave slope that can be predicted as it is recorded can be ignored and the predicted slope is recreated in playback. This is tricky business at best, but it works. The problem is that it only works at times when the waveform is simpler and can be predicted.

Another system for compressing audio involves the concept that a person can't hear something loud and soft at the same time. If a cannon is fired during the *1812 Overture*, for an instant no one can hear anything but the cannon. And the wave slope of the cannon is huge and therefore rather predictable. The violins are a complex waveform and any compression of them would be heard, but even if they stopped playing during the split second of the cannon fire, no one would hear the difference. So, compression can be increased during loud, low-frequency sounds.

It's also easy to interleave multichannels and save tremendous file size. Six-channel sound need not be six times as large as one channel. Four-channel LCRS Dolby is actually only two analog channels, yet these two channels contain all four channels discretely encoded into these two channels using nothing more than phase angle. In this case, any sound recorded only in the left channel is in fact a left channel sound. Any sound recorded in phase into both left and right channels is in fact a center channel sound and will be routed to the center speaker. Any sound in both right and left, where the right and left sounds are exactly 180° out of phase, is a surround channel sound. This is called multiplexing and it is possible when there are multiple signals. There are scores of surround formats and compression schemes, and all use compression and some use multiplexing to one degree or another.

Digital processing

Digital audio also can be processed while in the digital form. It can be pitch-shifted, time-shifted, delayed, filtered, expanded, compressed, and many other types of processing, all mathematically. Some of this processing is unique to the digital realm; others are models of analog processes. Many of these digital models are not true to the analog process, they claim to model. Yet many come very close and many are just as rich and interesting than their analog counterpart. However, there is still need to perform many analog processes to audio. If you want a sound to sound like it is playing inside of a big wood box, while it might be possible to build a digital filter that makes audio sound more or less like it was recorded in a wood box, the best and easiest way to get this effect is to rerecord the sound inside of a big wood box. The downside of this is that if you want to change the size of the box after hearing the sound at the mix, other than stopping, building a new box, and rerecording the sound, the sound can't be changed. If the sound is processed to sound like it is in a box, it can be changed at will.

Processing can be real-time or rendered. Real-time processing can be added to the data stream; the digital audio stream simply plays through the digital device or software plug-in. Rendered effects are applied to a digital audio file that is rendered into a new duplicate audio file. There are advantages and disadvantages to both. The advantages to rendered effects are that they are finished and self-contained. Assuming that the effect is good, the rendered effected audio is used like any other audio. The disadvantages are that the rendered effect cannot be easily altered and that there are some disadvantages to being self-contained. Normally, altering a rendered clip will require going back to the original media and redoing the effect and render. Also, the effect cannot be altered over time; the effect is applied equally to the entire clip. Because the media is self-contained, the effect cuts off at the end of the media. In the case of a delay or reverb, this can be clearly heard if the media cuts off sharply.

A processor attached to a channel of audio has none of these limitations but will be applied to the entire track and will therefore require management to avoid applying an effect to audio that should not be effected.

Digital audio has revolutionized audio production and postproduction. Capabilities have gone up while costs have plunged down. Complex edits can be cut in a fraction of the time and with fewer people. While there was originally an antidigital backlash, for the most part, the vast majority would say that the quality has also gone right through the roof. The systems are now becoming simpler, more reliable, and cropping up everywhere. No doubt someone is editing a film soundtrack in Podunk, New York, at this very moment.

Index

Printed and bound by CPI Group (UK) Ltd, Croydon, CR0 4YY

22/10/2024

01777530-0002